AMC PAMPHLET

AMCP 706-290

THIS IS A REPRINT WITHOUT CHANGE OF ORDP 20-290 DATED JULY 1959

RESEARCH AND DEVELOPMENT OF MATERIEL

ENGINEERING DESIGN HANDBOOK

WARHEADS-GENERAL (U)

Regraded UNCLASSIFIED
By Authority of DA 1575 HQ DARCOM,
DRCSS-I, Dtd 16 Mar 76

Signature *[signature]*   22 Mar 76 Date

HEADQUARTERS, U. S. ARMY MATERIEL COMMAND   JULY 1964

This material contains information affecting the national defense of the United States within the meaning of the Espionage Laws, Title 18, U.S.C., Sec 793 and 794, the transmission or revelation of which in any manner to an unauthorized person is prohibited by law.

GROUP 3
Downgraded at 12 year intervals; not automatically declassified.

~~CONFIDENTIAL~~

HEADQUARTERS
UNITED STATES ARMY MATERIEL COMMAND
WASHINGTON, D.C. 20315

31 July 1964

AMCP 706-290(C), <u>Warheads--General</u> (U), forming part of the Army Materiel Command Engineering Design Handbook Series, is published for the information and guidance of all concerned.

(AMCRD)

FOR THE COMMANDER:

OFFICIAL:

SELWYN D. SMITH, JR.
Major General, USA
Chief of Staff

R. O. DAVIDSON
Colonel, GS
Chief, Administrative Office

DISTRIBUTION: Special

UNCLASSIFIED

~~CONFIDENTIAL~~

PREFACE

The Engineering Design Handbook of the Army Materiel Command is a coordinated series of handbooks containing basic information and fundamental data useful in the design and development of Army materiel and systems. The Handbooks are authoritative reference books of practical information and quantitative facts helpful in the design and development of Army materiel so that it will meet the tactical and technical needs of the Armed Forces.

This handbook on Warheads--General presents information on the fundamental principles governing the design of warheads, with discussions of the mechanical and explosive arrangements which have been, or may be, used in the construction of warheads. More detailed and extensive treatment of specialized designs in warheads is contemplated in subsequent handbooks.

This handbook was prepared under the direction of the Engineering Handbook Office of Duke University, prime contractor to the Army Research Office-Durham. The material was prepared by Aircraft Armaments, Inc., under subcontract to the Engineering Handbook Office. Technical assistance was rendered by Picatinny Arsenal and the Ballistics Research Laboratories of Aberdeen Proving Ground. During the preparation of this handbook Government establishments were visited for much of the material used and for helpful discussions with many technical personnel.

Agencies of the Department of Defense, having need for Handbooks, may submit requisitions or official requests directly to Publications and Reproduction Agency, Letterkenny Army Depot, Chambersburg, Pennsylvania 17201. Contractors should submit such requisitions or requests to their contracting officers.

Comments and suggestions on this handbook are welcome and should be addressed to Army Research Office-Durham, Box CM, Duke Station, Durham, North Carolina 27706.

TABLE OF CONTENTS

Chapter Number	Chapter Title	Page
	Table of Contents	ii
	List of Illustrations	iii
	List of Tables	vii
	Glossary of Warhead Terms	viii
	Symbol Definitions	xi
1	Warhead Types	1
1-1	Introduction	1
1-2	Blast Warheads	1
1-3	Fragmentation Warheads	3
1-4	Discrete Rod Warheads	7
1-5	Continuous Rod Warheads	7
1-6	Cluster Warheads	9
1-7	Shaped Charge Warheads	10
1-8	Chemical and Biological Warheads	11
1-9	Incendiary Warheads	12
1-10	Leaflet Warheads	12
1-11	Inert and Exercise Warheads	13
2	Weapons System Concepts	30
2-1	Introduction	30
2-2	Scope	30
2-3	The Measure of the Cost of the Contribution	31
2-4	Size of the Weapons System Design Team	31
2-5	Application of Weapons System Concept to Warhead Design	32
3	Warhead Selection	33
3-1	Introduction	33
3-2	Classification of Targets	33
3-3	Warhead Selection Chart	34
3-4	Bibliography	35
4	Warhead Detail Design	36
4-1	General	36
4-2	Blast Warheads	36
4-3	Fragmentation Warheads	46
4-4	Discrete Rod Warheads	84
4-5	Continuous Rod Warheads	87
4-6	Cluster Warheads	98
4-7	Shaped Charge Warheads	111
4-8	Chemical and Biological Warheads	130
4-9	Characteristics of Service Warheads	138
5	Warhead Evaluation	139
5-1	Evaluation Principles	139
5-2	Fundamental Concepts	149
5-3	Approximate Evaluations	153
5-4	Evaluation Methods	154

6	Warhead Testing	167
6-1	Introduction	167
6-2	Planning of Test Program	167
6-3	Test Procedures and Techniques	170
6-4	Data Reduction and Interpretation	182
6-5	Test Facilities	184
6-6	Bibliography	188
	Appendix: Characteristics of High Explosives for Missile Warheads	189

LIST OF ILLUSTRATIONS

Figure Number	Title	Page
1-1	Typical Internal Blast Warhead With Penetration Nose	1
1-2	Typical External Blast Warhead	1
1-3	Blast Warhead - Aerial Target	2
1-4	Blast Warhead - Surface Target	2
1-5	Damage from Blast Warhead	3
1-6	Fragmentation Warhead - Aerial Target	4
1-7	Fragmentation Warhead - Surface Target	5
1-8	Typical Individual Fragment Shapes	6
1-9	Preformed Fragment Retention	6
1-10	Fire-Formed Fragment Casings	7
1-11	Fragment Orientation about Charge	7
1-12	Typical Fragment Warhead	8
1-13	Damage from Fragmentation Warhead	9
1-14	Discrete Rod Warhead	10
1-15	Typical Discrete Rods	11
1-16	Typical Discrete Rod Warhead	11
1-17	Damage from Discrete Rod Warhead	12
1-18	Continuous Rod Warhead	13
1-19	Typical Continuous Rod Warhead	13
1-20	Damage from Continuous Rod Warhead	14
1-21	Cluster Warhead	15
1-22	Submissile Shapes	15
1-23	Submissile Ejection	16
1-24	Skin Removal	16
1-25	Example of Cluster Warhead	16
1-26	Damage from Cluster Warhead	17
1-27	Shaped Charge Warhead - Aerial Target	18
1-28	Shaped Charge Warhead - Surface Target	19
1-29	Action of Shaped Charge Warhead	20
1-30	Typical Shaped Charge Warhead	20
1-31	Aircraft Damage from Shaped Charge Warhead	21
1-32	Armor Penetration from Shaped Charge Warhead	22
1-33	Chemical or Biological Warhead	23
1-34	Current Bomblet Shapes	24
1-35	Typical Biological Warhead	24

1-36	Incendiary Warhead	25
1-37	Typical Incendiary Bomblet	26
1-38	Typical Incendiary Warhead	26
1-39	Leaflet Warhead	27
1-40	Exercise Warhead - Aerial Target Drone	28
1-41	Inert Warhead	29
2-1	Utilization of Manpower and Natural Resources	30
2-2	Engineering Effort - Guided Missile System Development	31
4-1	Action of Internal and External Blast	37
4-2	Peak Pressure Vs. Scaled Distance at Various Atmospheric Pressures, 50/50 Pentolite Spherical Bare Charges	43
4-3	Scaled Impulse Vs. Scaled Distance at Various Atmospheric Pressures, 50/50 Pentolite Spherical Bare Charges	43
4-4	One Piece Fabrication	44
4-5	Multipiece Fabrication	44
4-6	Definition of Fragment Beam Width	47
4-7	Vector Addition of Fragment and Missile Velocities	47
4-8	C_D Vs. Mach Number for Various Fragment Types	48
4-9	Longitudinal Section of a Typical Fragmentation Warhead	51
4-10	Diagram for Derivation of Angle of Emission of Fragments	51
4-11	Distribution of Fragments about Nominal Ejection Direction	51
4-12	Graphical Solution of Optimum Beam Width	54
4-13a	Examples of the Effect of Warhead Shape on Fragment Beam Width	54
4-13b	Examples of the Effect of Warhead Shape on Fragment Beam Width	55
4-14	Volume Per Pound of Warhead Vs. Charge-to-Metal Ratio	57
4-15	Initial Static Fragment Velocity Vs. Charge-to-Metal Ratio	57
4-16	Warhead Effectiveness Vs. Fragment Size	60
4-17	Fragment Velocity and Size Optimization, Target: Piston Engine Fighter	60
4-18	Fragment Velocity and Size Optimization, Target: Piston Engine Fighter	60
4-19	Fragment Velocity and Size Optimization, Target: Piston Engine Fighter	60
4-20	Fragment Velocity and Size Optimization, Target: B-29 Aircraft with Fuel	61
4-21	Fragment Velocity and Size Optimization, Target: B-29 Aircraft with Fuel	61
4-22	Fragment Velocity and Size Optimization, Target: B-29 Aircraft with Fuel	61
4-23	Fragment Velocity and Size Optimization, Target: B-29 Aircraft with Fuel Invulnerable	62
4-24	Fragment Velocity and Size Optimization, Target: B-29 Aircraft with Fuel Invulnerable	62
4-25	Fragment Velocity and Size Optimization, Target: B-29 Aircraft with Fuel Invulnerable	62
4-26	Fragment Velocity and Size Optimization, Target: Single Engine Jet Fighter	62
4-27	Fragment Velocity and Size Optimization, Target: Single Engine Jet Fighter	63
4-28	Fragment Velocity and Size Optimization, Target: High Explosive Airborne Bomb	63
4-29	Fragment Velocity and Size Optimization, Target: High Explosive Airborne Bomb	63
4-30	Fragment Velocity and Size Optimization, Target: High Explosive Airborne Bomb	63
4-31	Fragment Velocity and Size Optimization, Target: High Explosive Airborne Torpedo	64

4-32	Fragment Velocity and Size Optimization, Target: High Explosive Airborne Torpedo	64
4-33	Fragment Velocity and Size Optimization, Target: High Explosive Airborne Torpedo	64
4-34	Velocity Ratio Vs. Range, Anti-Personnel Warhead	64
4-35	Vector Addition of Fragment and Missile Velocities, Anti-Personnel Warhead	65
4-36	Fragment and Spray Diagram: Unit-Radius Sphere for Burst at 60° ω of Warhead Designed for the Same ω	69
4-37	Fragment and Spray Diagram: Unit-Radius Sphere for Burst at 30° ω of Warhead Designed for 60° ω	70
4-38	Warhead Efficiency Vs. Warhead Inclination	81
4-39	Typical Use of a Fairing (Sparrow I, Mk 7 Mod 0, Warhead Shown)	82
4-40	Expected Number of Cuts Vs. Rod Length (Cylinder Half Severed)	86
4-41	Expected Number of Cuts Vs. Rod Length (3' x 8' Beam Half Severed)	86
4-42	Warhead Details (Discrete Rod Warhead)	86
4-43	Expansion of Rod Hoop	89
4-44	Rod Velocity Variation	89
4-45	A Warhead Design Producing Tangled and Broken Rods	89
4-46	Contoured Liner and Inert Build Up	89
4-47	Contoured Inert Build Up	91
4-48	Rod Scabbing	91
4-49a	Continuous Rod Warhead Designs	91
4-49b	Continuous Rod Warhead Designs	93
4-49c	Continuous Rod Warhead Designs	93
4-49d	Continuous Rod Warhead Designs	94
4-49e	Continuous Rod Warhead Designs	94
4-49f	Continuous Rod Warhead Designs	95
4-49g	Continuous Rod Warhead Designs	95
4-49h	Continuous Rod Warhead Designs	96
4-49i	Continuous Rod Warhead Designs	96
4-49j	Continuous Rod Warhead Designs	97
4-49k	Continuous Rod Warhead Designs	97
4-49l	Continuous Rod Warhead Designs	98
4-50	Resolution of Velocities	102
4-51	Unstabilized Submissiles - Typical Arrangement	108
4-52	Stabilized Submissiles - Typical Arrangement	112
4-53	Drag Tube and Drag Plate Stabilizer	113
4-54	Folding Fin Stabilizer	114
4-55	Drag Chute Stabilizer	114
4-56	Fixed Fin Stabilizer	114
4-57	Integral Ignition System	115
4-48	Central Ignition System	116
4-59	Piston Type Ejection	117
4-60	Blast Ejection - Intermediate Liner	117
4-61	Blast Ejection - Segmented Chamber	117
4-62	Blast Ejection - Convoluted Liner	117
4-63	Submissile Arrangement	118
4-64	Ejection Tube and Case	118
4-65	Gun Tube Ejection	118

4-66	Pressure/Length Curves	118
4-67	Maximum Allowable Chamber Pressure Vs. Gun Tube Geometry	118
4-68	Maximum Allowable Chamber Pressure Vs. Gun Tube Geometry	119
4-69	Maximum Allowable Chamber Pressure Vs. Gun Tube Geometry	119
4-70	Spherical Submissile Warhead	119
4-71	Charge-to-Metal Ratio Vs. Velocity for T-46 Cluster Warhead	119
4-72	T-46 Prototype Warhead	120
4-73	Typical Submissile Flight Pattern	120
4-74	Spherical Submissile	120
4-75	Types of Stabilization	120
4-76	Fin Stabilized Submissile	121
4-77	Fin Lock	121
4-78	Support Structure	121
4-79	Tube Analysis	121
4-80	Submissile Retention	121
4-81	Skin Removal Harness	122
4-82	Guillotine Installation	122
4-83	Guillotine Effect	122
4-84	Shaped Charge Nomenclature	123
4-85	Penetration Vs. Standoff; Cone Angle and Cone Thickness Against Concrete	124
4-86	Penetration Vs. Standoff and Cone Thickness Against Concrete	124
4-87	Penetration Vs. Standoff Against Mild Steel Target	125
4-88	Penetration Vs. Standoff Against Mild Steel Target	126
4-89	Penetration Vs. Standoff Against Mild Steel Target	127
4-90	Penetration Vs. Cone Angle Against Concrete	127
4-91a	Penetration Vs. Standoff Against Mild Steel Targets	127
4-91b	Penetration Vs. Standoff Against Mild Steel Targets	127
4-91c	Penetration Vs. Standoff Against Mild Steel Targets	128
4-91d	Penetration Vs. Standoff Against Mild Steel Targets	128
4-92	Reasonable Values of Cone Wall Thickness for Copper Cones	129
4-93	Penetration Vs. Cone Thickness and Cone Angle Against Concrete	129
4-94	Explosive Charge Wave Shaping	129
4-95	Bomblet Dispersion Patterns	131
4-96	Ejection Sequence, Spheres	132
4-97	Ejection Sequence, Fletners and Gliders	137
5-1	Attack with Perfect Guidance and Perfect Fuzing	142
5-2	Orientation of the Axes	142
5-3	Random Guidance Errors	143
5-4	The Normal Frequency Function	143
5-5	Area Under Normal Frequency Curve	143
5-6	The Normal Curve of Error	144
5-7	Orientation of Rectangular Target Area	144
5-8	Probability of a Hit, P_Δ, within a Circle of u Standard Errors	146
5-9	Cumulative Bivariate Normal Distribution (Over Circles of Radius R Centered at the Mean)	148
5-10	Random Fuzing Errors Combined with Random Guidance Errors	150
5-11	Random Warhead Bursts Around Target	156
5-12	Critical Distance, d, for Evaluation of Blast Warhead	157
5-13	Critical Distance, d, for Evaluation of Fragment Warhead	157

5-14	Geometry for Fragment Striking Velocity	158
5-15	Critical Distance, d, for Evaluation of Rod Warhead	158
5-16	Critical Distance, d, for Evaluation of Cluster Warhead	159
5-17	Critical Angle, θ, for Evaluation of Shaped Charge Warhead	159
5-18	External Blast Warhead Evaluation-Warhead Weight Variable	159
5-19	External Blast Warhead Evaluation-Target Type Variable	160
5-20	External Blast Warhead Evaluation-Engagement Altitude Variable	160
5-21	External Blast Warhead Evaluation-Standard Error of Guidance Variable	160
5-22	External Blast Warhead Evaluation-Kill Probability Intervals	160
5-23	Internal Blast Warhead Evaluation-Target Type Variable	161
5-24	Internal Blast Warhead Evaluation - Engagement Altitude Variable	161
5-25	Internal Blast Warhead Evaluation - Standard Error of Guidance Variable	161
5-26	Internal Blast Warhead Evaluation - Kill Probability Intervals	161
5-27	Fragment Warhead Evaluation - Warhead Weight Variable	162
5-28	Fragment Warhead Evaluation - Target Type Variable	162
5-29	Fragment Warhead Evaluation - Standard Error of Guidance Variable	162
5-30	Fragment Warhead Evaluation - Kill Probability Intervals	162
6-1	Fragment Lethality Test	170
6-2	Fragment Gun and Sabot	171
6-3	Screen Used for Measuring Velocity	173
6-4	Individual Rod Test	174
6-5	Complete Fragment Warhead Test	175
6-6	Submissile Ejection System Test	176
6-7	Skin Removal Test	177
6-8	Sled Test - Complete Cluster Warhead	178
6-9	Complete Shaped Charge Warhead Test	179
6-10	Blast Warhead Test	180
6-11	Complete Rod Warhead Test	184

LIST OF TABLES

3-1	Target Classification	33
3-2	Warhead Selection Chart	34
4-1	Characteristics of Explosives	39
4-2	Characteristics of HBX-1 and H-6	39
4-3	Penetration Capabilities of Penetration Case	42
4-4	One Piece Fabrication	44
4-5	Nose Spray Warhead Characteristics	68
4-6	Estimated Relative Fragment Production From Various Fragmentation Control Methods	78
4-7	Rough Numerical Comparisons of Various Fragmentation Control Methods	78
4-8	Summary Chart	99
4-9	Relative Penetration Capabilities of Various Liner Materials	125
4-10	Characteristics of Existing Service Warheads	134
5-1	The Areas Under the Normal Curve of Error (Included Between t and -t)	145
5-2	The Probabilities of a Hit, P_m, Within a Circle of Radius u Standard Errors	146
5-3	Cumulative Bivariate Normal Distribution Over Circles Centered at the Mean	147
6-1	Test Facility Selection Chart	183

GLOSSARY OF WARHEAD TERMS

Aim Point - That point at which the warhead would detonate if all component systems functioned perfectly.

Biological Warhead - A warhead containing organisms, which damages primarily by inflicting diseases.

Blast Warhead - A warhead containing a high explosive charge which, upon detonation, creates a blast wave that inflicts damage by either the positive or the negative pressure phase, or both.

Bomblet - One of the many containers of lethal agents included in a missile warhead.

Casing - The material which forms the outer shell of a warhead.

Chemical Warhead - A warhead containing chemical agents, which damages primarily by toxic effects.

Cluster Warhead - A warhead containing a group of submissiles or bomblets, together with an ejection system.

Conditional Kill Probability - The probability of inflicting specified damage provided the target is detected, the guidance system functions, the warhead is delivered to the target, and the fuzing system functions.

Continuous Rod Warhead - A warhead designed to emit an expanding metal hoop as the primary damaging agent.

Detonating Cord - A plastic-covered textile wrapper containing a core of explosive material.

Discrete Rod Warhead - A warhead designed so that metal rods are the primary damaging agent.

Dynamic Fragment Velocity - The velocity in free air of the fragments from a warhead in motion.

Ejection System - The system that is used in cluster warheads for dispersing submissiles.

Elevon - Combination elevator and aileron, controlling both roll and pitch.

Evaluation - Determination of warhead performance, often relative to the original requirements for which it was developed.

Exercise and Inert Warheads - Warheads designed to be used for training and systems operation checking purposes. Formerly known as practice and training warheads respectively.

External Blast Warhead - A warhead designed to cause damage by blast when detonated in the vicinity of the target.

Fairing - Sheet metal skin installed around the warhead to maintain the missile aerodynamic contour.

Fragment - Piece of metal scattered by the detonation of a warhead.

Fragment Beam Width - Angle covered by a useful density of fragments.

Fragment Density - Number of fragments per square foot at a given distance from the point of detonation.

Fragment Pattern - The arrangement of fragments after detonation.

Fragmentation Warhead - A warhead so designed that metal fragments emitted at high velocities are the primary damaging agent.

Fuze - A device designed to initiate detonation of a warhead at a specific time or position, under certain desired conditions.

Guidance Error - The shortest distance between the missile trajectory and the aim point.

Guidance System - A group of electronic and mechanical devices designed to direct a missile to a target.

Hard Target - A target that is relatively difficult to damage as required.

Implosion - A force tending to create inward collapse.

Incendiary Material - A substance capable of setting fire to the target.

Incendiary Warhead - A warhead containing incendiary material as the primary damaging agent.

Internal Blast Warhead - A blast warhead designed to detonate upon impact or after penetration of a target.

Leaflet Warhead - A warhead containing leaflets or pamphlets.

Lethal Distance - The maximum distance at which a specific warhead can inflict lethal damage on a specific target.

Lethality - A measure of the effectiveness of a warhead.

Miss Distance - The distance between the burst point and the center of gravity of the target.

Missile - A self-propelling pilotless weapon.

Overpressure - That air pressure greater than the ambient air pressure.

Practice Warhead - See "Exercise and Inert Warheads".

Proximity Fuze - An electronic fuze which senses the presence of a target and initiates the detonation of the warhead at a certain distance from the target.

S & A - Safety and arming device.

Shaped Charge Warhead - A warhead designed to emit a jet of minute, hyper-velocity metal pieces which act as the primary damaging agent.

Soft Target - A target which may be damaged with relative ease.

Standard Error of Guidance - A measure of the dispersion (linear standard deviation) of guidance error.

Submissile - An individual unit containing explosive or other active agent, which forms only part of a missile warhead.

Target - The object or group of objects which a missile is employed against for the purpose of inflicting damage.

Training Warhead - See "Exercise and Inert Warheads".

Warhead Compartment - That space in a missile which is allocated to the warhead.

SYMBOL DEFINITIONS

A	Average presented area of fragment, square inches.	M_B	Maximum bending moment between loads, foot-pounds.
A_B	Bore area, square inches.	M_r	Ratio of casing weight to charge weight in cylindrical section.
A_L	Lethal area, square feet.	N	Total number of missiles fired.
A_T	Target projected vulnerable area, square feet.	N_f	Total number of fragments.
C_D	Drag coefficient.	$N(M)$	Number of fragments of mass greater than M.
D	Diameter, feet or inches.	P	Peak pressure, psi.
D_R	Drag, pounds or poundals.	P_a	Probability of a hit between ± a.
$D(\theta_d)$	Dynamic fragment density or a given direction θ_d, number of fragments per steradian.	P_b	Probability of a hit between ± b.
		P_c	Probability that the missile system will launch the missile.
$D(\theta_s)$	Static fragment density, number of fragments per steradian.	P_d	Probability that the missile will deliver the warhead to the target.
E	Energy per unit mass of explosive.		
F	Force, pounds or poundals.	P_Δ	Probability of a hit within a circle of radius m.
H	Burst height, feet.		
I	Positive impulse, psi - milliseconds.	P_f	Probability that a fuzing system will function.
K	Inside radius/outside radius.	P_h	Probability of a hit.
L	Length, feet or inches.	P_k	Conditional kill probability.
M	Mass, grains, ounces or pounds.	P_o	Pressure, atmospheres or psi.
M'	Weight to charge ratio parameter	P_r	Probability of detecting and/or recognizing a target.
M_o	Mean fragment mass, grains, ounces or pounds.	P_s	Overall kill probability.
M_A	Maximum bending moment at loads, foot-pounds.	P_z	Probability of fuzing within any given z

Symbol	Definition	Symbol	Definition
R	Distance from explosion, feet.	V_R	Radial velocity, feet per second.
S	Burst point.	V_s	Resultant striking velocity of fragments, feet per second.
R_D	Distance traveled, feet.	V_t	Velocity of target, feet per second.
R_G	Radius of centroid of tubular cross section, inches.	V_T	Total allowable warhead volume, cubic inches.
R_L	Lethal distance, feet.	V_x	Velocity at any point, feet per second.
T	Thickness of armor, inches.	V_{xs}	Absolute fragment velocity at the target, feet per second.
T_A	Total tangential compression at loads, pounds.	$V(T)$	Vulnerability of target.
T_B	Maximum tangential compression between loads, pounds.	W	Weight of warhead, pounds.
V	Total warhead volume, cubic inches.	W'	Equivalent bare charge weight, pounds.
$\bar{V}$	Average velocity, feet per second.	$W_{D.W.}$	"dead" weight in warhead, pounds.
V_D	Velocity of detonation wave in the explosive, feet per second.	W_E	Weight of individual submissile ejection tube, pounds.
V_d	Initial dynamic fragment velocity, feet per second.	W_m	Weight of fragmenting metal, pounds.
$V_{D.W}$	"dead" volume, cubic inches.	W_n	Net weight, pounds.
V_f	Initial static fragment velocity, feet per second.	W_p	Weight of submissile, pounds.
V_g	Longitudinal component of fragment velocity, feet per second.	W_s	Weight of individual submissile support structure, pounds.
V_L	Lethal striking velocity, feet per second.	W_T	Total allowable warhead weight, pounds.
V_m	Missile velocity, feet per second.	W_x	Weight of individual submissile explosive, pounds.
V_n	Net volume, cubic inches.	c	Weight of explosive, pounds.
V_o	Initial relative velocity, feet per second.	c/m	Charge-to-metal ratio.
V_p	Submissile volume, cubic inches.	d	Significant distance in warhead evaluation, feet.
		$\dfrac{dV}{dt}$	Acceleration of a fragment along its

	path, feet per second².	z	Fuzing error, feet.
g	Acceleration due to gravity, ft/sec².	η	Warhead efficiency.
h	Altitude of engagement, feet.	θ	Angle between the missile and target trajectories, degrees.
l	Total travel distance, inches or feet.	θ_b	Beam width, degrees.
$l(m)$	Lethality of the missile.	θ_d	Angular direction, degrees.
m	Weight of the metal case, pounds.	θ_f	Half angle between forces, degrees.
m_p	Projectile mass, pounds.	θ_r	Angle of rod trajectories above or below the horizontal, degrees.
n	Number of submissiles.		
n_{xi}	Number of missiles which have an x component of error equal to x_i.	θ_s	Static angle of fragment ejection, degrees.
p_i	Probability of bursting at the i'th position.	ω	Angle of inclination of the missile with the ground, degrees.
p_{ki}	Probability of a kill given a burst from the i'th position.	$\psi(F)$	Frequency distribution of fuzing error.
		ϕ	Angle of fragment emission, degrees.
$p_v(x)$	Pressure producing velocity.	$\phi(G)$	Frequency distribution of guidance error.
r	Radius of submissile pattern, feet.	α	Constant, characteristic of explosives.
r_i	Inside radius, inches or feet.	β	A specific angle.
r_o	Outside radius, inches or feet.	$\phi(t)$	Normal frequency function.
t	Time, seconds.	$\phi(x)$	Frequency distribution of x.
x	Distance traveled, inches or feet.	$\phi(y)$	Frequency of distribution function of y.
$\ddot{x}$	Projectile acceleration, feet per second².	$\phi(z)$	Frequency distribution of z.
x_i	x component of guidance error for the i'th missile.	ρ_a	Air density, pounds per cubic foot or slugs per cubic foot, as applied.
x_s	Distance traveled to strike point, feet.	Δ	Guidance error of the missile, feet.
y_i	y component of guidance error for the i'th missile.	ρ_c	Density of charge, pounds per cubic inch.

ρ_m	Density of metal, pounds per cubic inch.
ρ_o	Air density at sea level, pounds per cubic foot or slugs per cubic foot, as applied.
$\rho(\Delta)$	Radial density function.
σ_G	Standard error of guidance.
σ_r	Radial compression, pounds.
σ_x	Standard deviation of x from the aim point, feet.
σ_y	Standard deviation of y from the aim point, feet.
σ_z	Standard error of fuzing, feet.
σ_θ	Tangential tension, pounds.
τ	Shear, pounds.

Subscripts:

f	Fragment.
i	Pertaining to the i'th missile.
n	Pertaining to the n'th missile.

Chapter 1
WARHEAD TYPES

1-1. INTRODUCTION

This chapter presents a general discussion of current warhead types. It consists of introductory material for use by those not familiar with warhead design art. The figures represent an artist's conception of the warhead for each type. These have been drawn to include the missile delivering the warhead, the warhead in operation and the target. They provide an overall concept of the warhead system in operation. Artistic license has been applied in some of these figures for clarity by showing the delivering missile intact after detonation of the warhead. This is not usually the case.

1-2. BLAST WARHEADS (Fig. 1-1 through 1-5)

Fundamentally, a blast warhead is high explosive installed in a container. Upon detonation, it creates a wave front of high positive pressure, followed immediately by a negative pressure. The wave moves radially outward at supersonic speed from the point of detonation. Primary damage occurs when a target is struck by the wave. Secondary damage usually results from flying debris. Blast warheads, installed in the body of a missile, are used against both air and surface targets.

Blast warheads are divided into two functional categories, internal and external blast. An internal blast warhead is designed to inflict damage when detonated upon impact with the target, or after penetration. When penetration of a hard target such as armor is required, the internal blast type is equipped with an armor piercing head. When penetration of a softer target such as aircraft structure is required, the warhead detonates upon impact and the extremely high pressures developed very close to the detonation point provide the means for entering the target. Since internal blast warheads must literally contact the target to be effective, they are normally used in missiles whose guidance systems are of the requisite accuracy. Extreme guidance accuracy is not required of missiles containing a number of internal blast submissiles, such as are carried by a cluster warhead. Fuzes for internal blast warheads are designed to detonate the warhead upon impact or very shortly thereafter.

The external blast warhead is designed to inflict damage when detonated near the target,

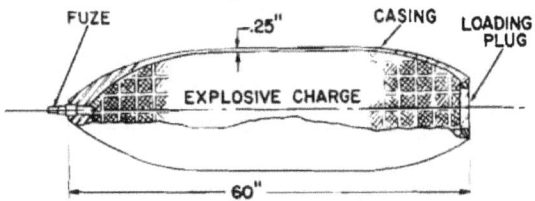

Figure 1-1. Typical Internal Blast Warhead With Penetration Nose

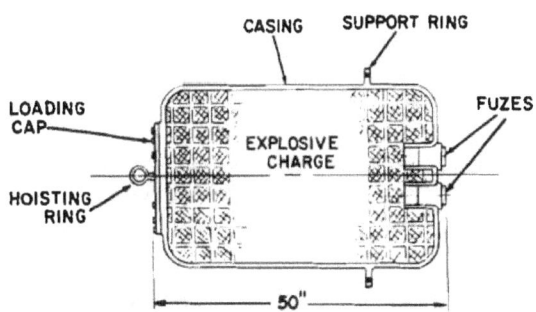

Figure 1-2. Typical External Blast Warhead

Figure 1-3. Blast Warhead - Aerial Target

Figure 1-4. Blast Warhead - Surface Target

U. S. Army Photo

View shows inspection of damage resulting from a blast warhead detonated beneath a Fighter Aircraft suspended in the air between two towers.

Figure 1-5. Damage from Blast Warhead

instead of upon striking the target. Consequently, it can be used in a missile whose guidance system provides less accuracy than that required for the internal blast type. Proximity type fuzes are used to detonate the warhead whenever it comes within lethal range of the target. This range depends on the size of the warhead, the target and the density of the air. It may be as low as 10 to 20 feet or as high as 150 to 200 feet.

Since the damage from both internal and external blast warheads is produced by a wave of high pressure air, the lethality of either warhead deteriorates significantly as the target altitude increases. Consequently, blast warheads, designed for use against air targets, are most effective when employed at low altitudes, say below 20,000 feet.

1-3. FRAGMENTATION WARHEADS

(Fig. 1-6 through 1-13)

A fragmentation warhead consists of an explosive charge surrounded by a wall of preformed metal fragments or a prescored or solid metal casing. Upon detonation, the fragments are propelled outward at velocities of from 6,000 to 10,000 feet per second by explosive forces. Generally slow-acting damage is inflicted on the target when it is struck by the fragments. Also, disruptive blast damage is common from fragmentation warheads. Fragmentation warheads, installed in the body of missiles are used against both air and surface targets, the surface targets most often being personnel. Their lethality against bomber targets depends upon the damage inflicted on a plurality of multiple components, such as

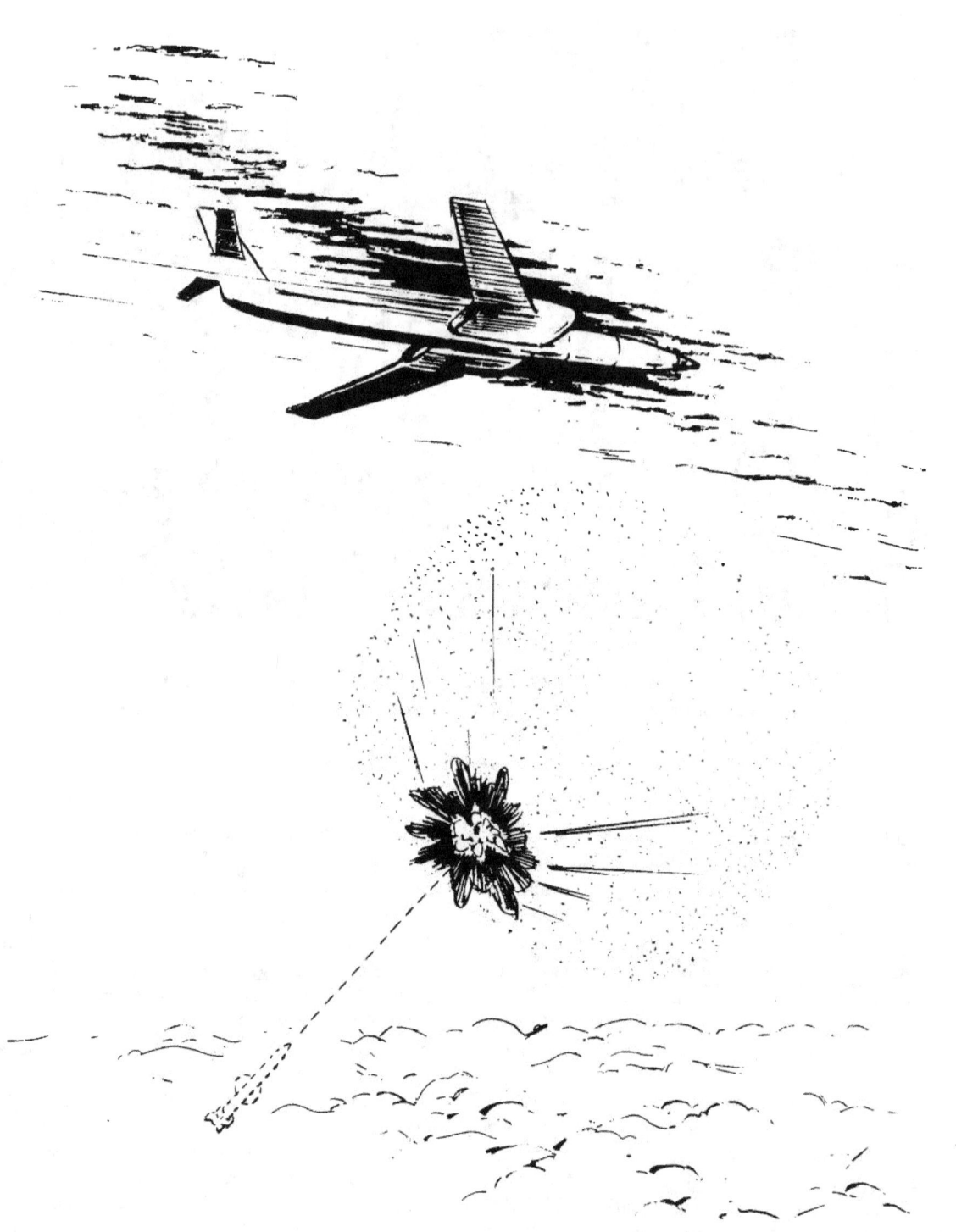

Figure 1-6. Fragmentation Warhead - Aerial Target

Figure 1-7. Fragmentation Warhead - Surface Target

engines, fuel lines, controls, instruments, hydraulic lines, and crew members.

The weight and shape of the individual fragments depends on the particular intended application of the warhead. Design fragment weights may vary from below .014 ounces (6.0 grains) to over 0.5 ounces (220 grains). Fragment shapes in past and current use include steel spheres, cubes, rods, wires, and aerodynamically stable configurations. (See Figure 1-8.) These shapes are either preformed or fire-formed. Preformed fragments are formed into their final shape before detonation of the explosive charge. They are mechanically held in their proper orientation around the charge by placing them in a fragment chamber, and either cementing them in place with adhesives or imbedding them in a plastic or frangible substance.

Fire-formed fragments take on their final individual shape during detonation of the explosive charge. Prior to detonation they are components of a fragment casing which surrounds the explosive charge. This casing is scored or notched in such a manner that it will break up upon detonation of the charge into individual fragments of the desired shape.

The pattern that the fragments form as a

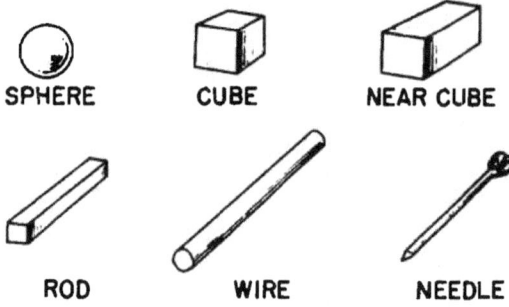

Figure 1-8. Typical Individual Fragment Shapes

group as they are propelled outward from the warhead by the explosive forces is primarily dependent upon their orientation around the charge prior to detonation. The patterns resulting from the orientations shown in Figure 1-11 are intuitively apparent. The casing shape selected for a particular warhead depends primarily on the guidance accuracy of the missile delivering the warhead, the size of the target and the density of the fragment beam on the target that is necessary to do lethal damage to the target.

Fragmentation warheads are designed to be detonated near, rather than upon striking, the target. Proximity-type fuzes are normally used and the guidance accuracy of the missile delivering the warhead need not produce a direct hit to insure lethality. In some instances, warhead detonation may be initiated by command from the ground. Fragmentation warheads used against air targets at high altitude may be lethal when detonated as far as 200 feet away. The lethal distance is primarily dependent upon the target and the fragment size and velocity at the time it strikes the target, since it must have sufficient kinetic energy to penetrate. Fragment striking velocity is a function of the target velocity, initial fragment velocity immediately after detonation and the aerodynamic drag forces which slow down the fragment during its flight to the target. The initial velocity is dependent upon the amount of explosive charge relative to the metal in the casing, and the weight and shape of the warhead. The drag forces depend on the shape of the

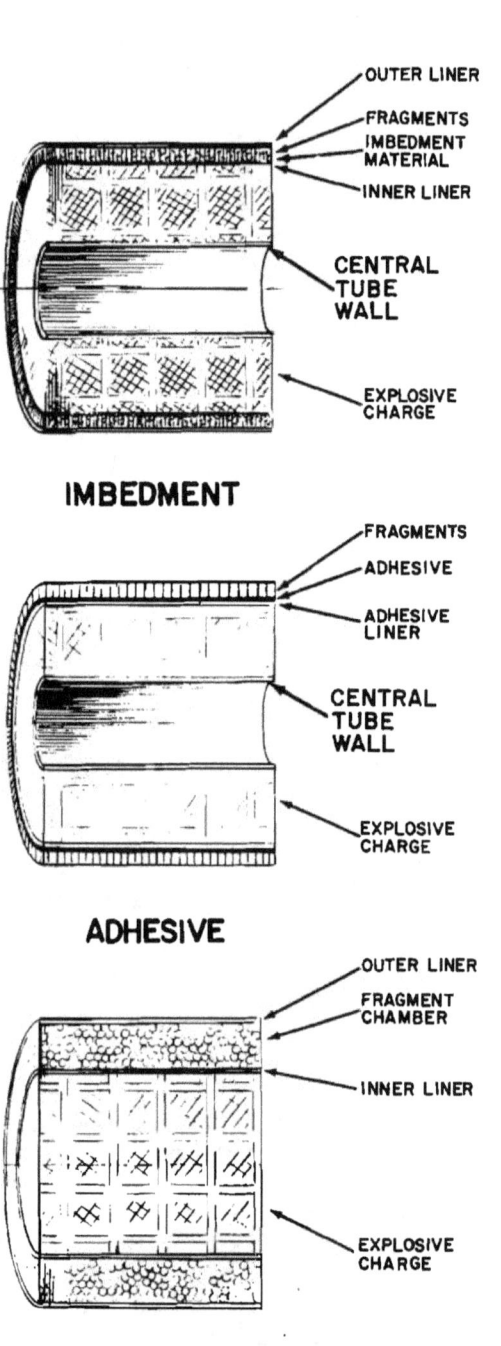

Figure 1-9. Preformed Fragment Retention

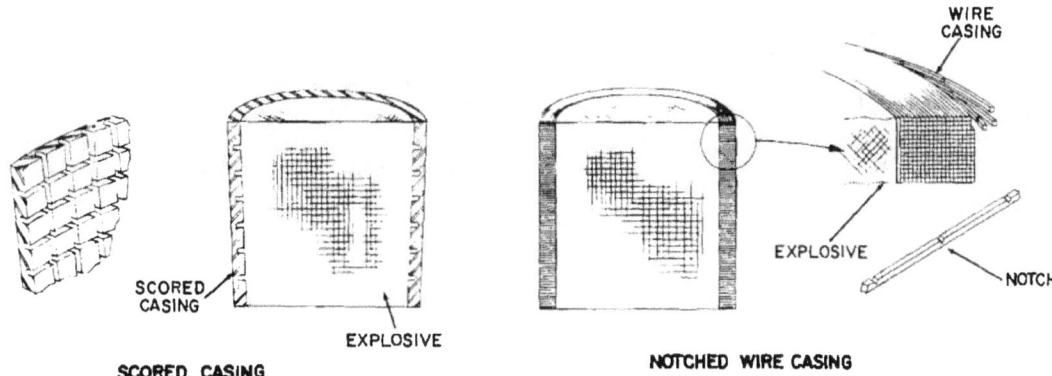

Figure 1-10. Fire-Formed Fragment Casings

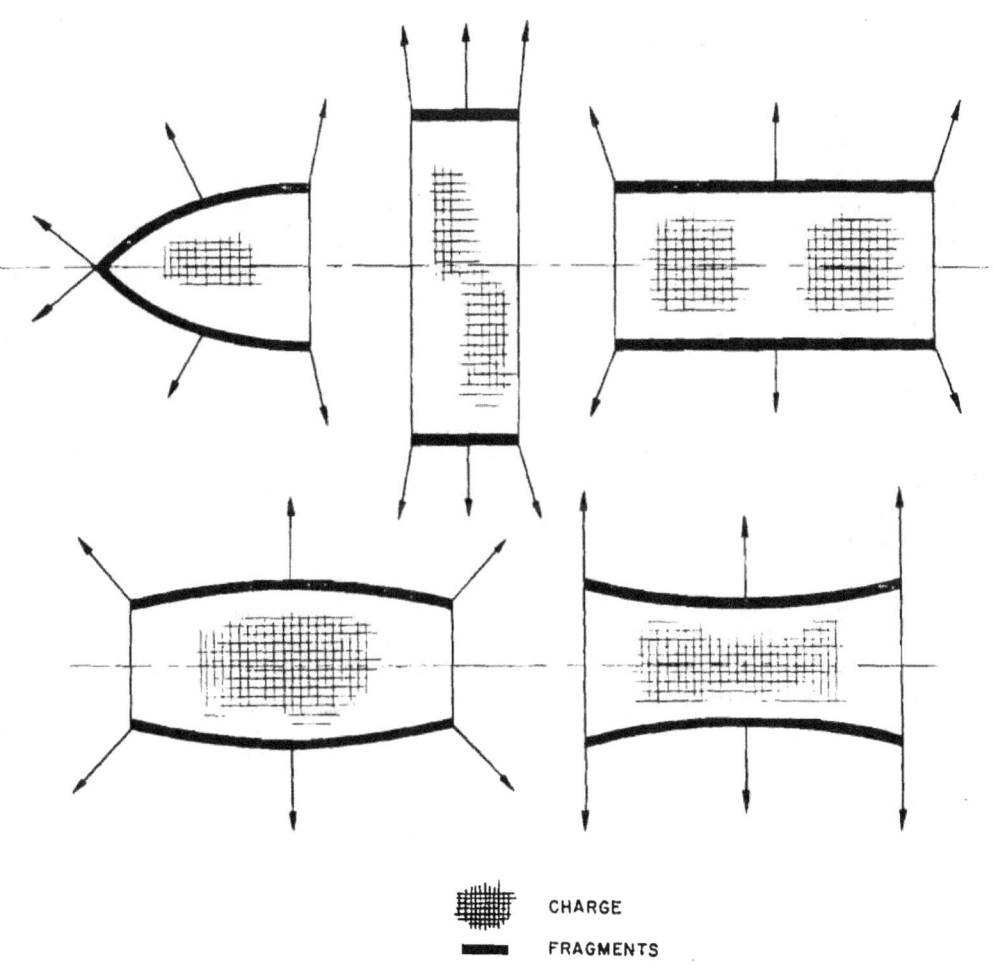

Figure 1-11. Fragment Orientation about Charge

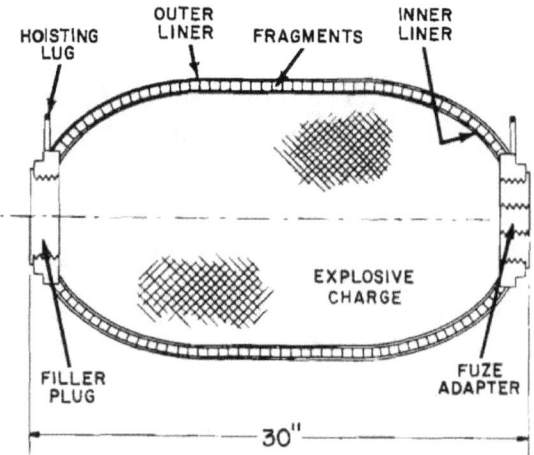

Figure 1-12. Typical Fragment Warhead

fragment and the density of the air. Consequently, preformed aerodynamically stable fragment shapes slow down less during their flight and are lethal at greater distances than random fragments produced by casing breakup. All fragment warheads are lethal to greater distances at higher altitudes due to the reduced drag of the rarer atmosphere.

1-4. DISCRETE ROD WARHEADS

A discrete rod warhead is similar to a fragmentation warhead (reference subchapter 1-3). It consists of an explosive charge surrounded by individual metal rods. Upon detonation, the rods are propelled outward at velocities of from 4,000 to 6,000 feet per second by the explosive forces and inflict damage on the target as they strike it. Discrete rod warheads, installed in the body of missiles, are used primarily against air targets. Their lethality stems from their ability to cut through and thereby critically weaken primary aircraft structure and aircraft system components.

The rods generally vary in thickness and length from 1/4 inch by 20 inches to 3/4 inches by 40 inches. Their cross-sectional shape may be circular, square, or trapezoidal. They are oriented about the explosive charge in one or more layers.

Discrete rod warheads are most effective against high altitude airborne targets similarly to fragmentation warheads by virtue of reduced slow down at high altitude during their flight to the target after detonation of the explosive charge. They have been displaced now to a great extent by continuous rod warheads discussed in subchapter 1-5.

1-5. CONTINUOUS ROD WARHEADS

(Fig. 1-18 through 1-20)

A continuous rod warhead consists of an explosive charge surrounded by a series of metal rods. Each rod is welded at one end to the rod adjacent to it on one side and is welded at the other end to the rod adjacent to it on the other side. Upon detonation of the explosive charge the rod welded assembly is propelled outward at velocities of from 3,000 to 5,000 feet per second or greater. As it moves outward from the point of detonation it forms a continuous and expanding hoop which eventually breaks up into several pieces as the hoop circumference approaches and exceeds the summation of the rod lengths. Continuous rod warheads, mounted in the body of missiles are employed primarily against air targets. They are lethal by virtue of their ability to critically weaken major structural components by a cutting action. One of the distinguishing characteristics of a continuous rod warhead is its ability to accomplish a "quick kill".

The thickness and length of the rods have varied from 3/16 inch by 10 inches to 5/8 inch by 40 inches. Their cross-sectional shape may be circular, square or trapezoidal. The maximum diameter of the expanded hoop depends of course on the number and length of rods and has varied between 30 and 200 feet. Since the rods are most effective before the hoop breaks up, the warhead is most lethal when detonated so that the hoop strikes the target before breakup.

The cutting ability of the rods is a function of the rod hoop weight and its velocity. Here again then, they are most effective when employed against air targets at high altitude since the rarer air at altitude causes less slow down of the hoop.

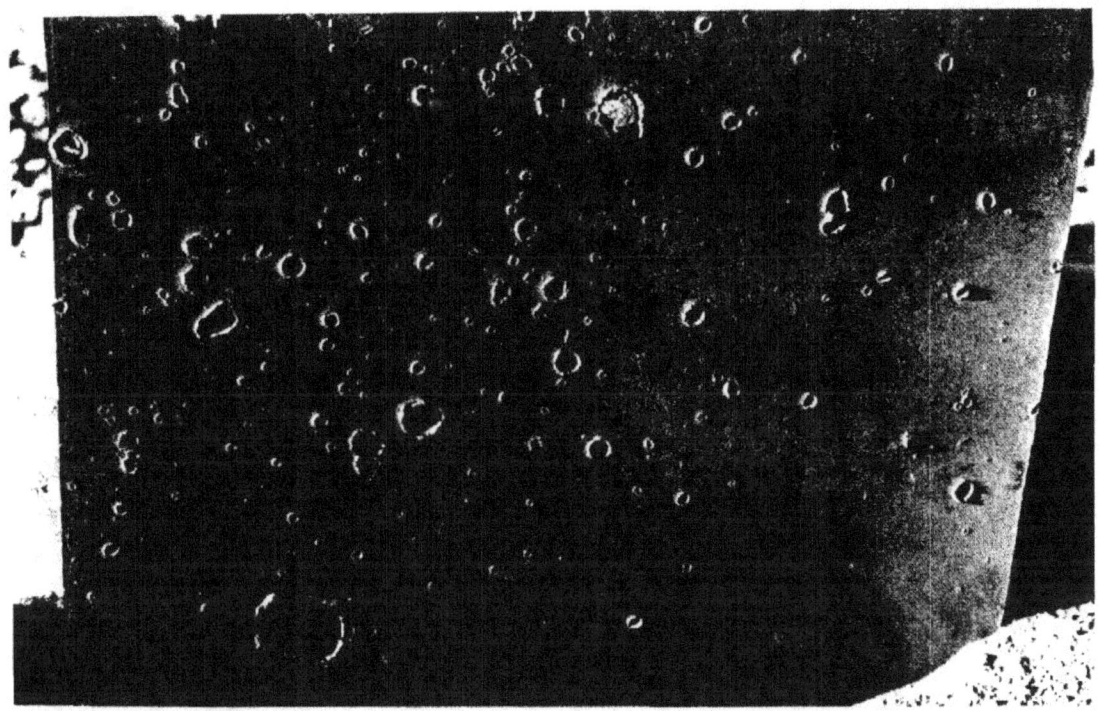

Aerojet Photo 955548

Secondary damage on Bikini gage protector plate, 60 feet to the rear of the T45 warhead, caused by fragmentation of the rear plate of the warhead. Controlled fragmentation is not provided in this area. Larger holes are from fragments; smaller holes are from covering material or from sand or grit.

Figure 1-13. Damage from Fragmentation Warhead

1-6. CLUSTER WARHEADS
(Fig 1-21 through 1-26)

Chemical, biological and incendiary warheads are types of cluster warheads which contain bomblets that are ejected from the warhead by aerodynamic forces after removal of the missile skin from around the warhead compartment. These warheads are discussed individually in subchapters 1-8 and 1-9, and the discussion in this section is limited to explosive-type cluster warheads.

A cluster warhead consists of a number of submissiles mounted in the warhead on individual ejection devices or surrounding an ejection device. Each submissile contains an explosive charge. Upon detonation of the warhead, the ejection device is actuated and the submissiles are propelled outwardly at velocities of from 100 to 500 feet per second. Damage is inflicted on the target as the explosive charge in the submissile is detonated upon striking the target. Cluster warheads, installed in the body of missiles, are used primarily against air targets. Their lethality is derived from the ability of one or a few submissiles to destroy a major component of the target or to inflict critical structural damage.

The number, weight, and shape of the individual submissiles depend primarily upon the weight and space allocated to the warhead in the missile. The number of submissiles may be as low as 10 or as high as several hundred.

Figure 1-14. Discrete Rod Warhead

Their individual weight is usually of the order of 3 to 5 pounds. The shape is dependent upon missile warhead compartment packaging considerations and upon whether an aerodynamically stable or unstable submissile is desired. Stability is obtained through fins or drag producing devices. Unstable shapes include spheres, cubes, and near-cubes.

One example of an ejection device consists of explosive actuated guns, one for each submissile. These consist of annularly displaced and radially directed tubes which slide into a close fitting cavity in the submissile. Simultaneous detonation of the propelling charge in the tubes shoots the submissiles outward from the missile.

Cluster warheads are in some instances designed for installation in the missile in such a manner that the outside of the submissiles forms the exterior surface of the missile body. In this case, the submissiles are accurately shaped and fitted so as to provide an aerodynamically acceptable surface prior to detonation of the warhead. In other cases, the warhead is housed within the missile skin. When this is done, the skin is usually removed by explosive means just prior to ejecting the submissiles so that the skin will not impede their ejection.

1-7. SHAPED CHARGE WARHEADS
(Fig. 1-27 through 1-32)

A shaped charge warhead has an axially symmetric high-order explosive arranged in a specific geometry with, generally, a detonating point located on the axis at one end of the charge and a symmetrically placed lined cavity at the other. In many cases the liner of the cavity is in the form of a cone with apex toward the detonator, but many other cavity shapes have been used. The principal characteristic of a shaped charge is that the shock wave in the explosive compresses some of the liner material into a high velocity stream called a jet. The forward end of the jet attains a velocity of from 16,000 to 20,000 feet per second while the aft end of the jet and the remaining liner material (called the "slug") have a forward velocity of about 1500 feet per second. Thus, if the material of the liner is sufficiently ductile and if there is sufficient space, the liner will

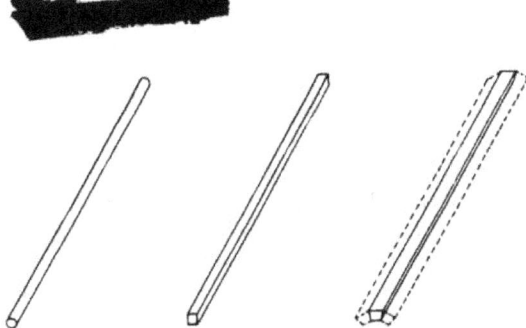

Figure 1-15. Typical Discrete Rods

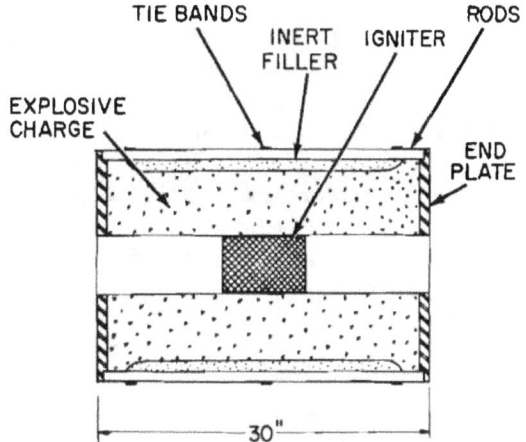

Figure 1-16. Typical Discrete Rod Warhead

draw out into a very long thin jet of extraordinary penetrating ability. The distance between the charge and the surface to be attacked is called the "standoff" and, depending on liner material and other parameters, there is generally an optimum standoff for greatest penetration. This distance, however, is seldom achieved in use.

Against armored targets, the damage inflicted by a shaped charge attack arises from the ability of the jet to penetrate large thicknesses of material and from the production of "spall" from the exit side of the surface attacked. Against aircraft and missile structure, where standoffs are large, the jet is usually broken up. Penetration of thin skins is therefore effected over a larger area and additional important damage (referred to as "vaporific") arises due to shock and blast effects in semi-enclosed structural spaces.

The most significant aspect of shaped charges lies in the cavity and liner. Liner shape, thickness and material are important variables. As examples, one may cite the use of copper cone liners .08 inches thick with 40° apex angles for use in the penetration of tank armor by 3.5 inch diameter shaped charge warheads. Against aircraft, aluminum liners 0.5 inches thick with 90° cone apex angles are typical for 8 inch diameter warheads.

Shaped charge warheads of weights stated later herein and detonated 1 to 2 feet from the surface of a target, can penetrate 1 to 4 feet of steel armor, or 15 to 20 feet of concrete. Large ones can be effective against aircraft when detonated 100 to 200 feet away. Since, in either case, the jet must impinge on the target, the missile guidance system must be such that the warhead is pointing toward or slightly leading the target when detonated.

A specialized application of the shaped charge phenomenon is the Misznay-Schardin effect in which the principle is used to form and propel fragments generated in the warhead casing. By shaping the explosive charge or by confining the sides of the charge, fragments can be ejected from certain portions in various beam widths. The fragment beam width is also governed by the L/D of the charge, and the charge-to-metal ratio.

Another type of warhead related to the shaped charge type is one known as the "squash-head" or high explosive plastic (HEP) warhead. However, since this type is easily defeated by the use of a soft layer in the target armor, it is not generally considered for use in missile warheads.

1-8. CHEMICAL AND BIOLOGICAL WARHEADS

(Fig. 1-33 through 1-35)

A chemical or biological warhead usually consists of a container housing a relatively large number of smaller containers known as bomblets. Upon release from the warhead, the bomblets, containing chemical or biological agents, are dispersed over a wide area on the ground. When the bomblet strikes the ground, the agent is released. Damage is inflicted upon targets such as personnel, animals and crops by contamination of the target or the air around the target in the region where the bomblets fall.

A wide variety of bomblet shapes have been studied and tested in an effort to obtain a shape which lends itself to efficient packaging in the warhead compartment and which will consistently disperse itself uniformly over large areas when large numbers of bomblets are released from the missile. Shapes which currently show the most promise are the ribbed sphere, the Fletner, and the glider. (See Figure 1-34.) Bomblets usually weigh between 4 and 10 pounds. As many as 500 bomblets are packaged in a single warhead. Dispersal areas on the ground may vary between 1 and 10 miles in diameter. Average ground distance between individual bomblets may vary from 50 to 300 feet.

Chemical and biological agents are not always dispersed by the use of individual bomblets in the warheads. Massive warheads consisting of one or a few larger containers for the agent may be jettisoned with or without a parachute from the missile warhead.

1-9. INCENDIARY WARHEADS
(Fig. 1-36 through 1-38)

Fundamentally, an incendiary warhead is a container for incendiary material. Incendiary material of a highly flammable nature is placed in small bomblets which are packaged in the warhead. When released from the warhead the bomblets fall to earth over a dispersed target area. Damage is inflicted on combustible targets when the incendiary material from the bomblets starts a large number of fires.

Because there have been no requirements established by the Department of Defense in recent years for incendiary warheads, development work has been limited to warheads using existing incendiary bomblets originally developed for use in clustered bombs.

Incendiary materials are separated into two basic categories, the intense type which burns at very high temperatures over a small area and the scatter type which burns at a lower temperature and is scattered over a wide area. Bomblets for the intense type contain a

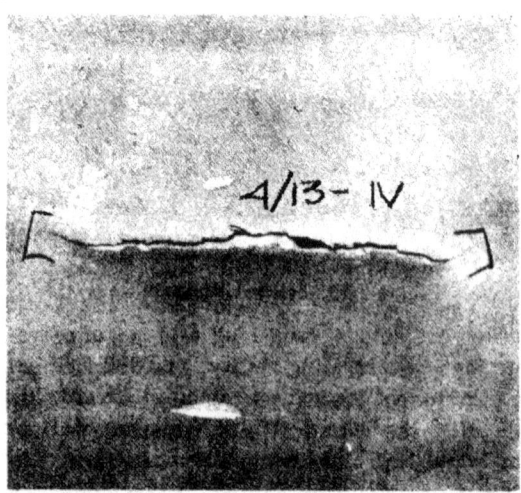

U. S. Navy Photo NP/9-48774

Fired through 3/4" thick 24S-T4 aluminum plate. View shows length of spall on back side of plate.

Figure 1-17. Damage from Discrete Rod Warhead

cast iron nose so that the bomblet penetrates the target prior to igniting. (See Figure 1-37.) For the scatter type, the bomblets are equipped with a small explosive charge which ignites and disperses the incendiary material over an area from 50 to 100 feet in diameter when detonated as the bomblet strikes the target. Suitable incendiary materials include thermite, white phosphorous, napalm and thickened mixtures of inflammable fuels.

The bomblet ejection system in the warhead consists of a means of removing the warhead skin. This is usually accomplished by inserting detonating cord between the skin and its supporting structural members. Detonation of the cord severs and blasts the skin away from the warhead, whereupon the bomblets are thrown out of the warhead by aerodynamic and gravitational forces.

1-10. LEAFLET WARHEADS
(Fig. 1-39)

A leaflet warhead consists of a container for housing leaflets or booklets. The leaflets are released from the warhead compartment

Figure 1-18. Continuous Rod Warhead

and fall to earth over a widely dispersed area Damage is inflicted through the demoralizing effect of the written material on the leaflets.

1-11. INERT AND EXERCISE WARHEADS

(Fig. 1-40 and 1-41)

Inert and exercise warheads are not used directly against the enemy, but rather are used to train personnel and for checking the operation of weapon systems and their components. An exercise warhead, formerly known as a practice warhead is a warhead which simulates the shape and weight of the tactical warhead. It is usually loaded with instruments which record or telemeter data on the performance of the weapon system components and operators during operation of the system. It is used against simulated targets on the surface and against target drones in the air. For use against air targets, it usually includes miss distance instrumentation which measures and records how close the delivering missile comes to the target during the practice flight. An exercise warhead may also contain, in lieu of instrumentation, pyrotechnic materials and small amounts of high explosive or spotting charges, and be tactically fuzed to provide realism.

An inert warhead, formerly known as a training warhead, simulates the shape, size,

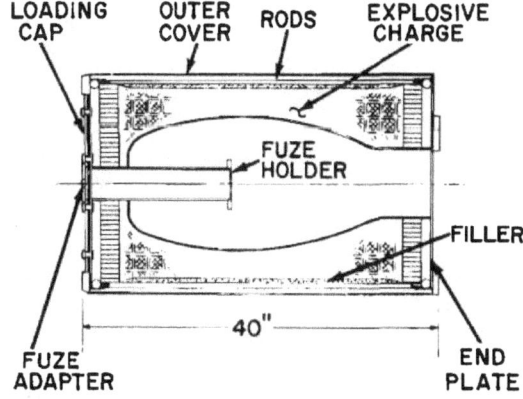

Figure 1-19. Typical Continuous Rod Warhead

New Mexico Institute of Mining and Technology Photo

*Figure 1-20. Damage from Continuous Rod Warhead
(Fired against aircraft skin panels)*

weight, support and handling provisions, external electrical, hydraulic, or pneumatic receptacles, and in general all components of the actual warhead which have an influence on the operations carried out by the weapon system ground crews. Such an inert warhead is not designed to be flown in the missile but instead is used for ground checkout and training of the weapon system operating personnel. It sometimes includes instrumentation to measure the accuracy and speed with which the ground crews carry out their particular functions.

Figure 1-21. Cluster Warhead

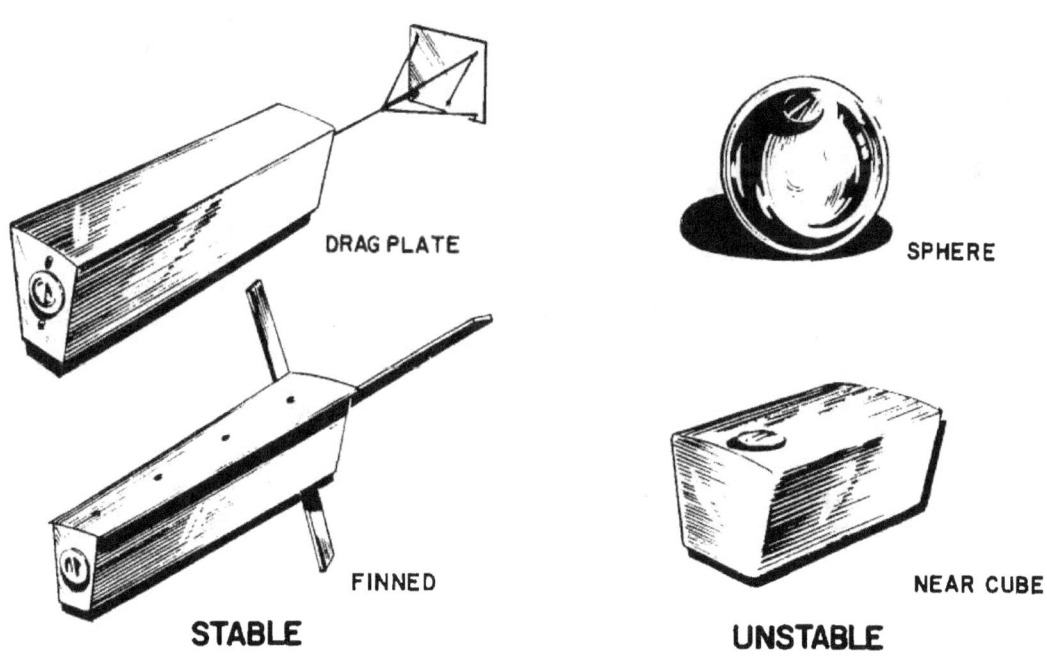

Figure 1-22. Submissile Shapes

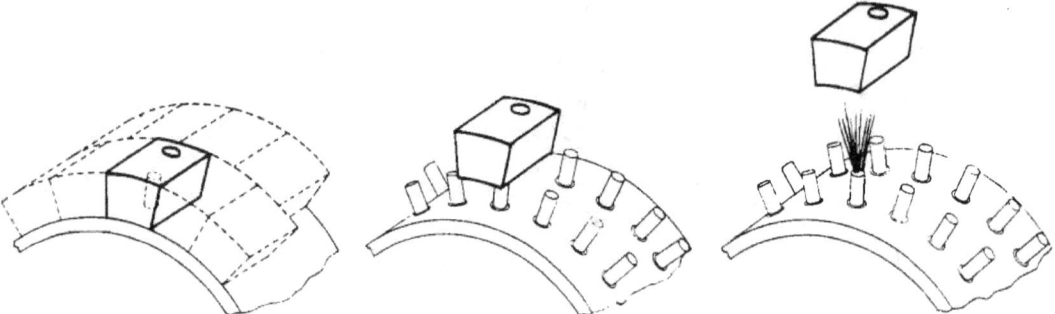

Figure 1-23. Submissile Ejection Gun Tube Method

Figure 1-24. Skin Removal

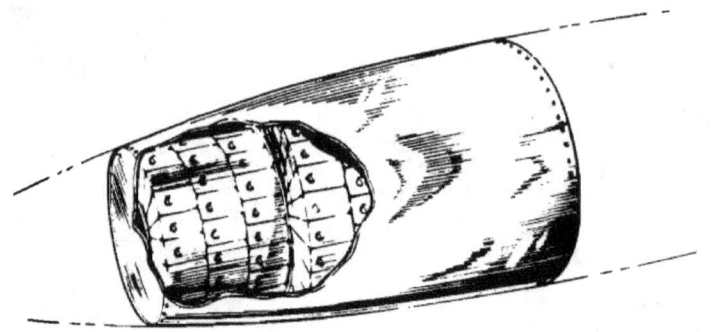

Figure 1-25. Example of Cluster Warhead

Entrance side of wing panel after being damaged by high explosive unstabilized submissile after impact with the target.

Figure 1-26. Damage from Cluster Warhead

Figure 1-27. Shaped Charge Warhead - Aerial Target

Figure 1-28. Shaped Charge Warhead-Surface Target

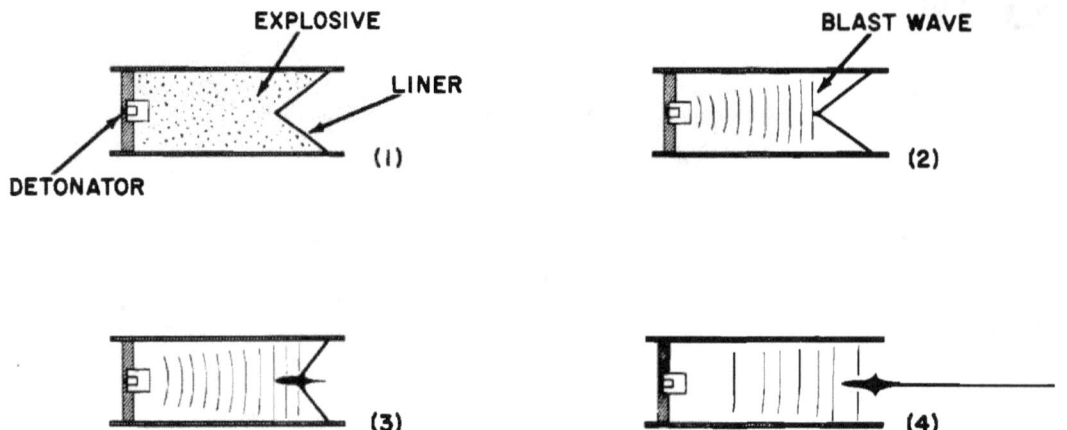

Figure 1-29. Action of Shaped Charge Warhead

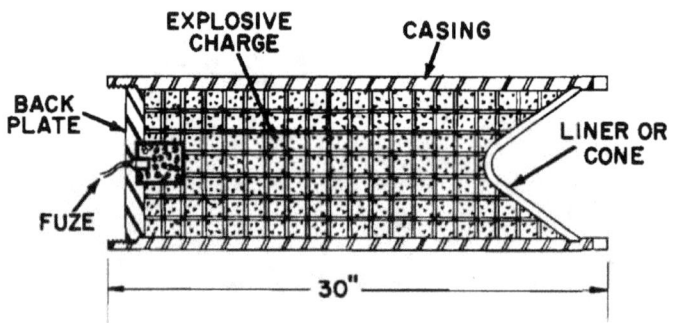

Figure 1-30. Typical Shaped Charge Warhead

U.S. Navy Photo NP/45-38575

View shows jet impact area of F6F-5K aircraft struck by aluminum coned charge at average jet velocity of over 14,200 fps for ~100 foot standoff.

Figure 1-31. Aircraft Damage from Shaped Charge Warhead

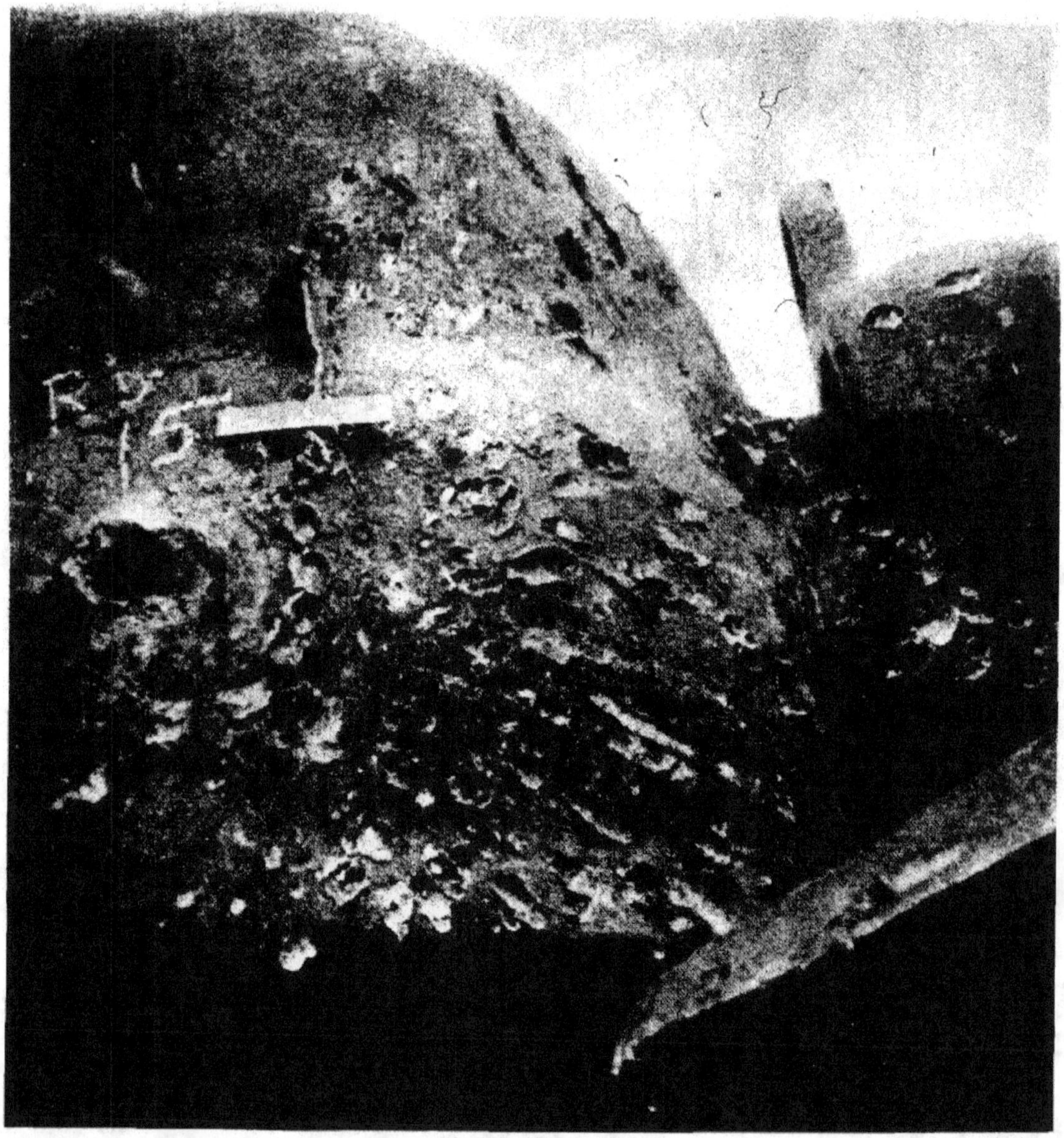

U. S. Army Photo

View shows impact location of 90 mm HEAT T108E40 projectile fired against a T26 Pershing type tank. Entrance hole is approximately 2-1/2 in. x 1-1/2 in.

Figure 1-32. Armor Penetration from Shaped Charge Warhead

Figure 1-33. Chemical or Biological Warhead

SPHERE

FLETNER

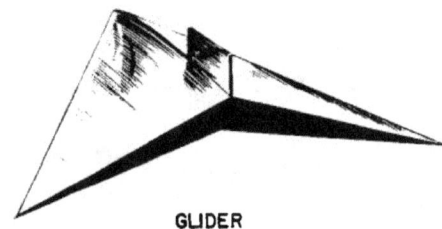

GLIDER

Figure 1-34. Current Bomblet Shapes

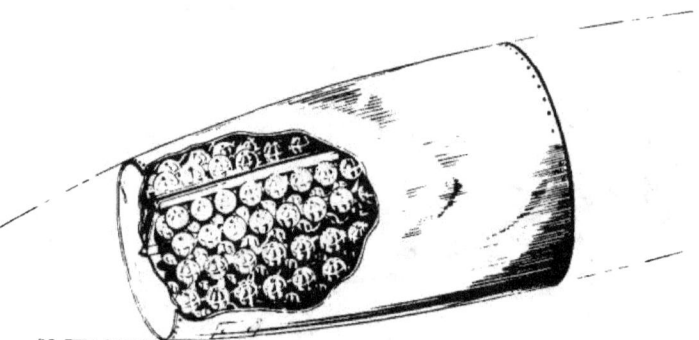

Figure 1-35. Typical Biological Warhead

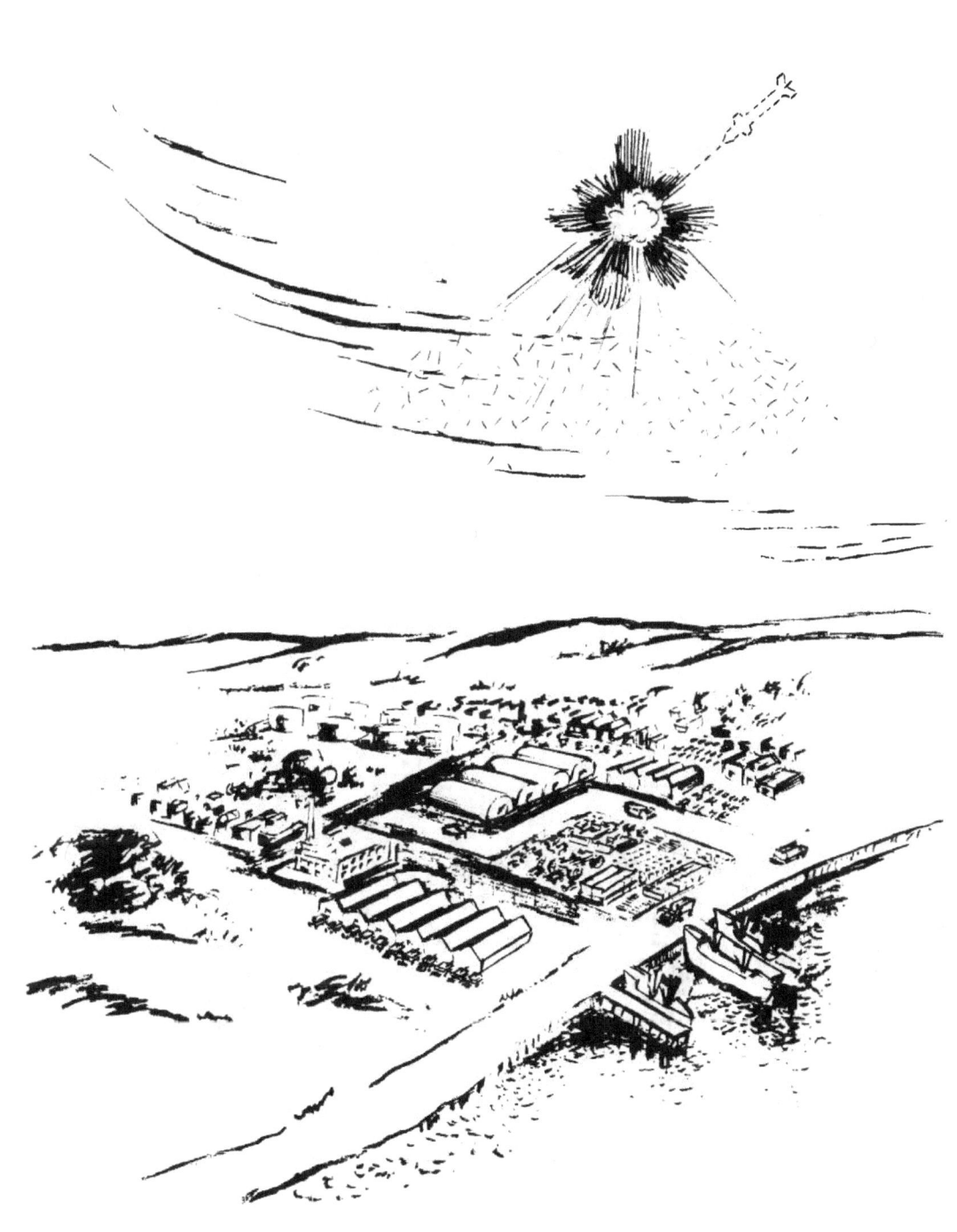

Figure 1-36. Incendiary Warhead

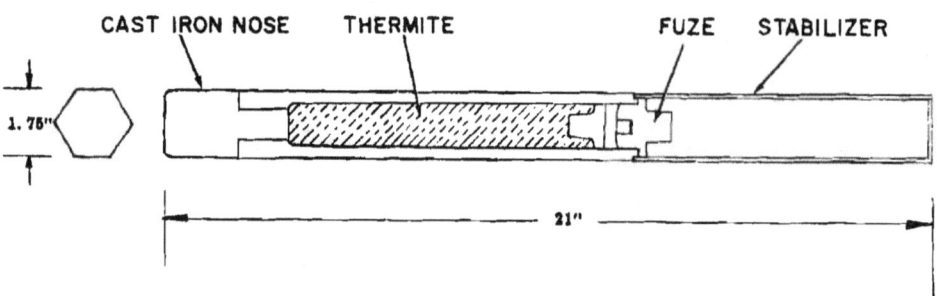

Figure 1-37. Typical Incendiary Bomblet

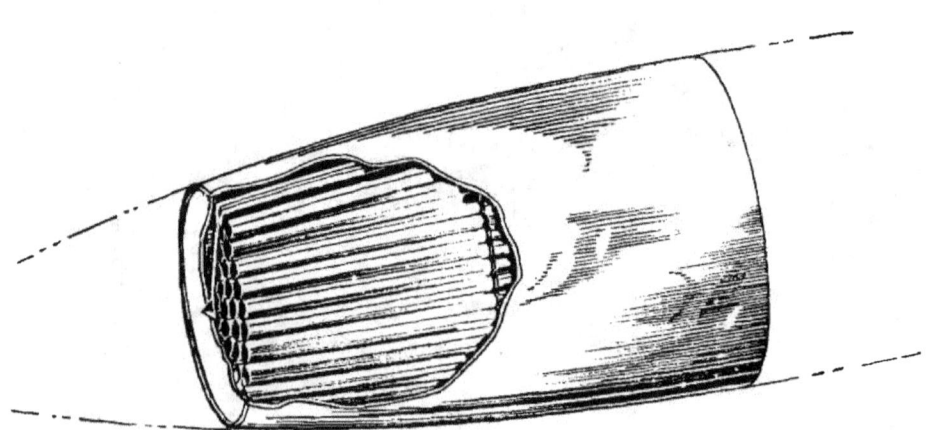

Figure 1-38. Typical Incendiary Warhead

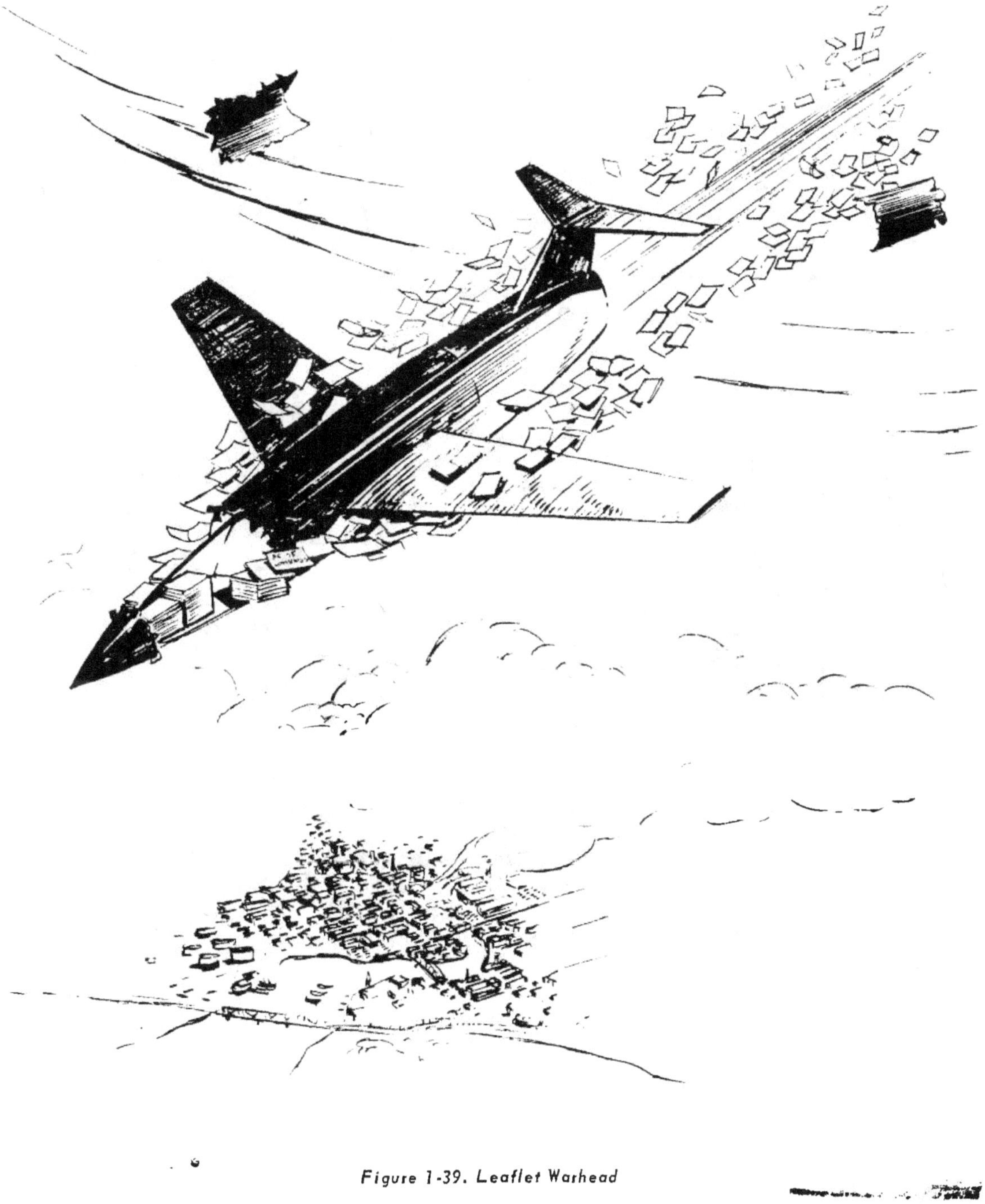

Figure 1-39. Leaflet Warhead

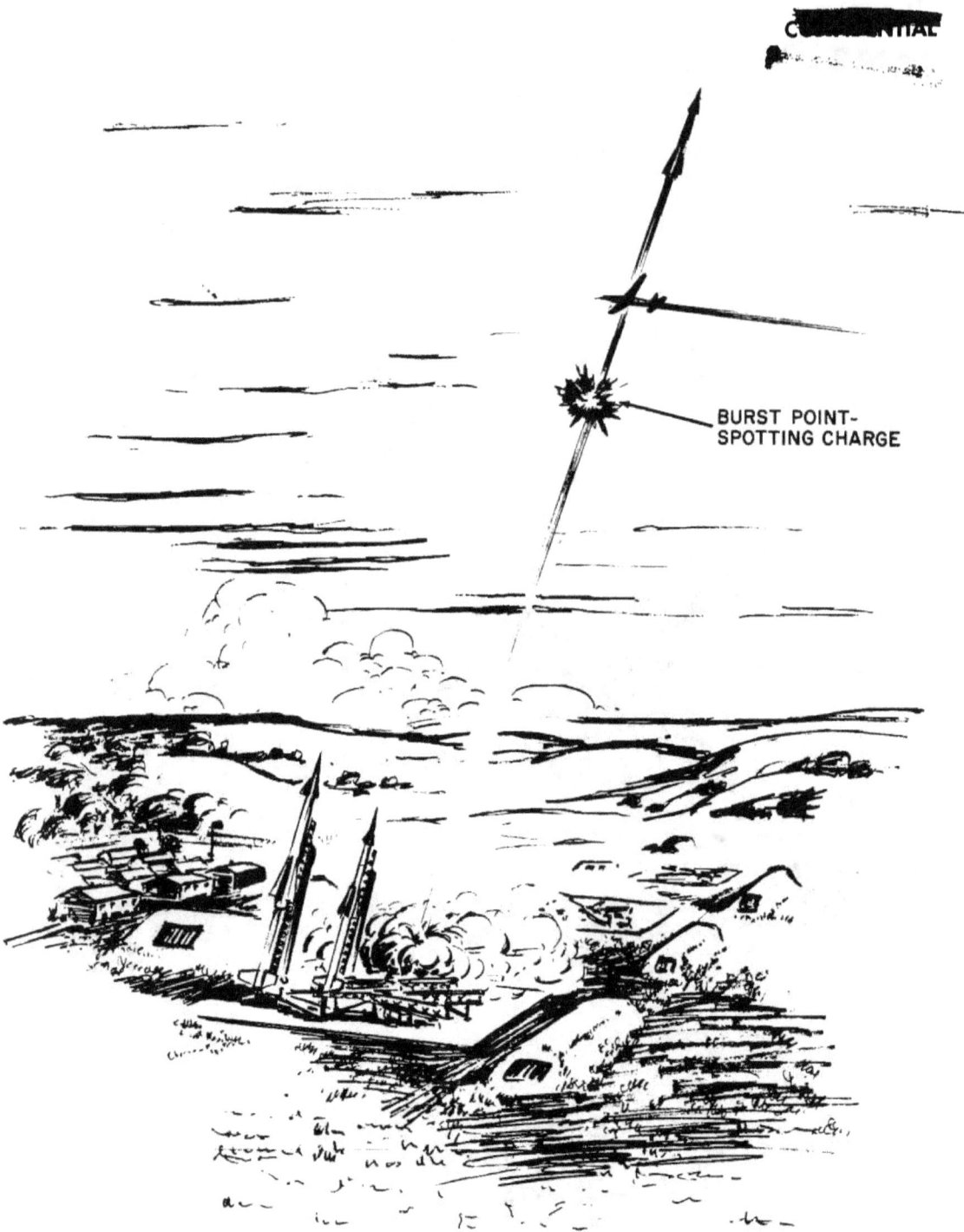

Figure 1-40. Exercise Warhead - Aerial Target Drone

Figure 1-41. Inert Warhead

Chapter 2
WEAPONS SYSTEM CONCEPTS

2-1. INTRODUCTION

The weapons system concept is a philosophy applied to the design of a multi-component system wherein each component of the system is so designed that its contribution to the complete system is a maximum when operating in conjunction with the other components of the system. This philosophy of design has long been applied in sound engineering practice. The increase in the complexity of modern weapons during recent years, along with the attendant increase in the size of the engineering staff required to design these weapons, brought forth a name for the philosophy of weapons system design. A discussion of the concept is included here since in practically every instance the warhead designer is working as a member of a weapons system team. It is important, then, that he appreciate the scope, principles, and the method of applying, weapons system design concepts.

2-2. SCOPE

The Department of Defense applies weapons system concepts in its broadest sense when it so utilizes the manpower and natural resources of this country that the contribution of each segment to the security of the United States is a maximum when operating in conjunction with all other segments. Weapons system concepts are then applied within the Department of Defense in a narrower sense, when, for example, an assignment is given to a weapons system team to develop a system for Air Defense. This air defense system might conceivably include four or five different means for destroying the attackers; one of which might be a surface-to-air missile system. This system is designed for maximum contribution to air defense by applying weapons system concepts in a still narrower sense. Furthermore, the warhead designer developing the warhead for this ground-to-air system designs his warhead for a maximum contribution to the system and thereby applies the concept. In fact,

PROCUREMENT	LOGISTICS	OPERATIONS

RESEARCH	STORAGE	TRAINING
DEVELOPMENT	SHIPMENT	TEST
PROTOTYPE MANUFACTURE	SUPPLY	EVALUATION
TEST	MAINTENANCE	DOCTRINE
PRODUCTION DESIGN	OVERHAUL	OFFENSIVE UTILIZATION
TOOLING AND PROCUREMENT	REPAIR	DEFENSIVE UTILIZATION
PRODUCTION MANUFACTURE	SPARES	ASSESSMENT

Figure 2-1. Utilization of Manpower and Natural Resources.

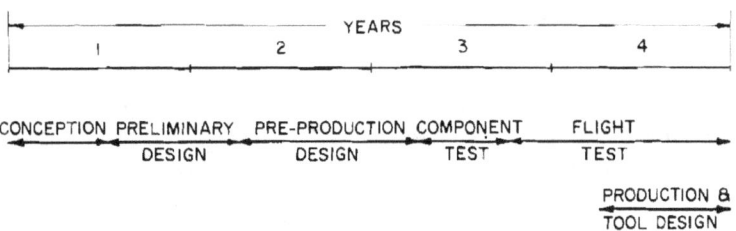

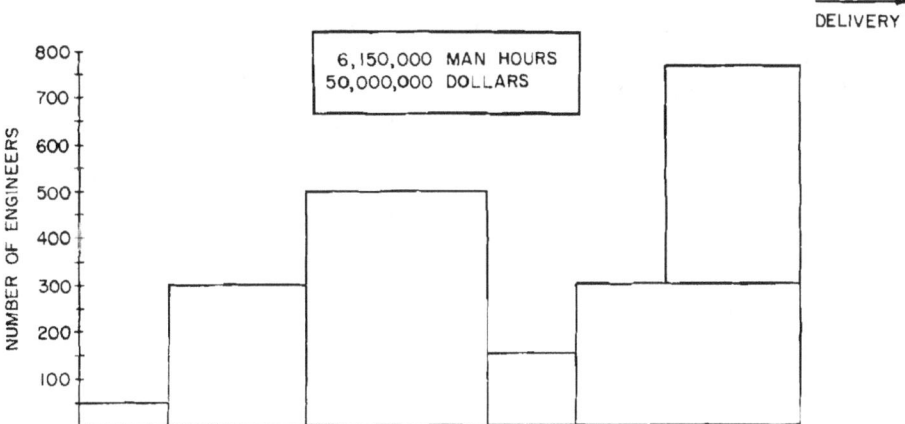

Figure 2-2. Engineering Effort - Guided Missile System Development

the detailed design of the various components of the warhead are developed so as to contribute to the maximum effectiveness of the warhead. Thus, it may be seen that the concept of weapons system design may be applied on a scope as broad as the operation of the Department of Defense or as narrow as the detailed design of a warhead component.

2-3. THE MEASURE OF THE COST OF THE CONTRIBUTION

The most difficult aspect of practicing the concept of weapons system design is to decide where to stop maximizing the contribution of a component. To make this decision, one must measure the contribution relative to the effort required to produce it. This effort, in the last analysis, is always measured in terms of manpower and natural resources. The much used terms of light weight, ease of maintenance, simplicity, reliability, low cost, strategic materials, and a host of other terms, all refer to conservation of manpower and natural resources. These are the true measures of cost, and therefore the optimum weapons system is one which obtains its objectives with the least overall expenditure of these two commodities. A breakdown of the utilization of these commodities in bringing a weapons system to bear on the enemy is shown graphically in Figure 2-1.

2-4. SIZE OF THE WEAPONS SYSTEM DESIGN TEAM

The size of a weapons system design team obviously varies widely in proportion to the scope of the problem. The missile warhead designer will most often be working as a member of a design team dealing with a complete missile system. The diagram shown in Figure 2-2 graphically represents the distribution of engineering effort for a typical missile system development. It is presented here to give the designer an understanding of how his work fits in with that of the other members of the team.

2-5. APPLICATION OF WEAPONS SYSTEM CONCEPT TO WARHEAD DESIGN

This pamphlet is written and arranged on the premise that the warhead designer will be working as a member of a weapons system team. As such, he will be responsible for all design and development work directly related to the warhead. His responsibilities will include warhead-type selection, warhead evaluation, detailed design in coordination with the missile and fuze designers, and warhead tests. Basic data are included in the pamphlet to assist the warhead designer in carrying out these responsibilities.

The warhead designer needs data relating to the complete missile system in order to carry out his function as the warhead member of the missile system team. These data are listed as follows. The pamphlet is written on the basis that this specific information is provided in each case for which a warhead design is required.

Information Supplied to the Warhead Designer
1. Allowable warhead weight
2. Warhead compartment size and shape
3. Missile velocity
4. Standard error of guidance
5. Size and shape of target
6. Target velocity
7. Target vulnerability
8. Target engagement altitude
9. Target engagement aspect
10. Warhead installation information
11. Missile environment during handling and flight

Chapter 3
WARHEAD SELECTION

3-1. INTRODUCTION

For a particular missile, the warhead designer may be required to select a warhead which will make the maximum contribution to the complete missile weapons system. If this is the case, in order to be certain that he has selected the proper type, he should actually carry out the detailed design of several types and then carefully evaluate each one against the other to determine the optimum. Preliminary design data for all warhead types are provided in Chapter 4. Chapter 5 provides basic information for evaluating the warhead types against various targets. This chapter (Chapter 3) presents data to aid the designer in selecting the warhead types for initial detailed design and evaluation.

The initial selection is based on the weight allotted for the warhead by the missile system designer, the standard error of guidance of the missile guidance system, and on the target or targets specified. This data, used in conjunction with the selection chart presented in subchapter 3-3, will permit the warhead designer to select one or two warhead types for optimization. If the chart does not indicate a unique type, the types selected must be designed in detail and evaluated.

3-2. CLASSIFICATION OF TARGETS

Targets are classified into eight types in the selection chart. These classifications are shown in Table 3-1.

These classifications have been chosen first so as to distinguish between aerial and surface targets. Aerial targets are further classified as high level or low level; a low level target is considered as being below 20,000 feet, a high level target as being above 20,000 feet. Surface targets are further subdivided into concentrated targets, dispersed targets and unprotected personnel targets. A concentrated target is one which requires a direct hit or very near miss to do damage and it can be disabled with one hit. A dispersed target is spread out over a large area and requires many hits to disable it. Personnel are in a separate class. Surface targets are further classified as hard and soft. A hard target is one which has been specifically designed and constructed to withstand attack, while a soft

TABLE 3-1

Target Classification

Classification	Title
HAA	High Altitude Aerial
LAA	Low Altitude Aerial
CHS	Concentrated Hard Surface
CSS	Concentrated Soft Surface
DHS	Dispersed Hard Surface
DSS	Dispersed Soft Surface
UPS	Unprotected Personnel on Surface
CPS	Partly Covered Personnel on Surface

3-3. WARHEAD SELECTION CHART

Warhead selections shown in Table 3-2 are tentative, and should be used as a very rough guide only.

Table 3-2 Warhead Selection Chart

TARGET CLASSIFICATION	STANDARD ERROR OF GUIDANCE (FEET)	WARHEAD WEIGHT IN POUNDS				
		25 or Less	25 to 100	100 to 300	300 to 600	Over 600
HAA High Altitude Aerial	Less than 30	SC* EB	SC* CR* FR	FR* CR* CL	FR* CL*	FR* CL
	30 to 60	SC EB	CR* FR	FR* CR* CL	FR* CL*	FR* CL
	More than 60		CR FR	CL* FR CR	FR* CL*	FR* CL
LAA Low Altitude Aerial	Less than 30	SC* EB	EB* SC*	EB* CL CR	EB* CL	EB* CL
	30 to 60	EB SC	EB* CR	EB* CL* CR	EB* CL*	EB* CL
	More than 60		CR EB	CL EB CR	CL* EB	EB* CL
CHS Concentrated Hard Surface	Less than 30	SC*	SC*	SC* IB	IB*	IB* EB
	30 to 60	SC	SC*	SC* IB	IB*	IB* EB
	More than 60			IB SC	IB	IB* EB
CSS Concentrated Soft Surface	Less than 30	FR* IB*	FR* IB*	FR* IB* EB	FR* IB* EB	FR* IB* EB
	30 to 60	FR*	FR* DR IB	FR* EB CL	FR* EB* CL	FR* EB*
	More than 60	FR*	FR* DR	FR* CL	FR* EB CL	FR* EB*
DHS Dispersed Hard Surface	Less than 30		SC	SC*	SC* IB	IB*
	30 to 60		SC	SC*	SC* IB	IB*
	More than 60			SC	SC	IB*
DSS Dispersed Soft Surface	Less than 30	FR*	FR* DR	FR* CL* DR	FR* CL* EB	FR* EB* CL
	30 to 60	FR*	FR* DR	FR* CL* DR	FR* CL* EB	FR* EB* CL
	More than 60	FR*	FR* DR	FR* CL* DR	FR* CL* EB	FR* EB* CL
UPS Unprotected Personnel on Surface	Less than 30	FR*	FR*	FR CL	FR CL	FR CL
	30 to 60	FR*	FR*	FR CL	FR CL	FR CL
	More than 60	FR*	FR*	FR CL	FR CL	FR CL
CPS Partly Covered Personnel on Surface	Less than 30	FR*	FR*	FR CL	FR CL	FR CL
	30 to 60	FR*	FR*	FR CL	FR CL	FR CL
	More than 60	FR*	FR	FR CL	FR CL	FR CL

CODE: IB - Internal Blast   FR - Fragment   CR - Continuous Rod   SC - Shaped Charge
       EB - External Blast   CL - Cluster   DR - Discrete Rod

*More likely selections.

target has not, although it may be military equipment.

Examples of the various classification of targets follows.

- HAA Turbojet aircraft, turboprop aircraft, reciprocating engine aircraft, turbojet missiles, ramjet missiles, rocket powered missiles.
- LAA Same as HAA plus helicopters, lighter-than-air craft.
- CHS Concrete pill boxes, bunkers, armored vehicles, single fortifications, tunnels or causeways, concrete dams, battleships, destroyers, large caliber gun emplacements, concrete bridges.
- CSS Trucks, locomotives, transport ships, tankers, landing craft, individual aircraft on ground, individual industrial buildings, wooden bridges.
- DHS Submarine pens, steel mills, underground industrial plants, Naval shipyards.
- DSS Large industrial complexes, railroad marshalling yards, airports, oil refineries, ammunition dumps, supply areas, highways.
- UPS Infantry troops in the field or in encampments.
- CPS Partly covered troops on the surface of the terrain, e.g. in trenches.

3-4. BIBLIOGRAPHY

(1) "Principles of Guided Missile Design; Operations Research; Armament", Grayson Merrill and Harold Goldberg, D. Van Nostrand Company, Inc., Princeton, 1956.

(2) "New Weapons for Air Warfare, Science in World War II", J. C. Boyce, Little, Brown, 1947.

(3) "Elementary Comparison of Antiaircraft Warhead Types", Herbert K. Weiss, BRL Memo 631, ASTIA AD-7624, Nov. 1952.

(4) "Optimization of Warhead and Fuzing Parameters", Rand Corp., RM-349, Mar. 1950.

(5) "Weapons Selection for Air Targets", Directorate of Intelligence (USAF), ASTIA AD-38743, Jan. 1954.

(6) "Tactical Analysis of Surface-to-Air Guided Missile Systems", C. F. Meyer, R. P. Rich and others, Johns Hopkins Univ. Report No. TB-166-1, ASTIA AD-22820, Nov. 1953.

(7) "Effectiveness of Warheads for Guided Missiles Used Against Aircraft", Ed S. Smith, BRL Memo. Report 507, Mar. 1950.

(8) "Effectiveness of Existing and Development Weapons in the Intermediate Antiaircraft Role", F. G. King and F. Q. Barnett, BRL Memo. Report 528, Nov. 1950.

(9) "Effectiveness of Missile Warheads Against High-Speed Air Targets", Ed S. Smith, BRL Memo. Report 528, Dec. 1954.

(10) "1500 lb. Anti-Personnel Warhead for the Honest John Rocket", Ed S. Smith, A. K. Eittreim and W. L. Stubbs, BRL Memo. Report 779.

Chapter 4
WARHEAD DETAIL DESIGN

4-1. GENERAL

This chapter presents the data needed to effect the complete detail design of all warhead types. A step by step procedure is set forth for each warhead type except chemical and biological which are presented in general narrative form. A summary of design data required and a list of the data required by the fuze designer is presented at the end of each subchapter.

The attention of the designer is again invited to the fact brought out in Chapter 2 that certain data are generally required to allow the warhead designer to effect the detail design of the warhead. The check list of these required data is repeated here for emphasis and convenient reference.

Information Supplied to the Warhead Designer
1. Allowable warhead weight
2. Warhead compartment size and shape
3. Missile velocity
4. Standard error of guidance
5. Size and shape of target
6. Target velocity
7. Target vulnerability
8. Target engagement altitude
9. Target engagement aspect
10. Warhead installation information
11. Missile environment during handling and flight

4-2. BLAST WARHEADS

4-2.1. Detail Design Steps

Step
Number   Detail Design Step
1. Decide Function of the Warhead Case
2. Investigate Compatibility of Weight and Space Allocated
3. Design, Installation and Handling Provisions
4. Make Strength Analysis
5. Provide for Loading Explosive
6. Provide for Installation of Fuze
7. Select Method of Fabrication
8. Prepare Summary of Fuzing Requirements
9. Prepare Summary of Design Data

The exact order of the design procedure may vary depending upon the viewpoint of the designer and, even more, on the military requirements which often fix certain parameters in advance.

4-2.2. Detail Design Data (Fig. 4-1 through Fig. 4-5; Tables 4-1 through 4-3)

Function of the Warhead Case and Related Blast Effects The function of the warhead case depends upon whether the warhead is designed to be detonated inside or outside of the target envelope, that is internal or external blast. The case for an internal blast warhead must function not only as a container for the explosive charge but also as a means for penetrating the target. The case for an external blast warhead serves only as a container for the charge.

Internal blast within a structure produces an overpressure inside of the structure. When this overpressure is of sufficient magnitude and duration, the target structure will fail due to the explosive action of the blast acting on the structure. This explosive action is such that the outer structure of the target is blown out and away from the target. The remaining structure is weakened to such an extent that it

INTERNAL

EXTERNAL

Figure 4-1. Action of Internal and External Blast

is unable to withstand the structural loads and collapse of the target structure occurs. See Figure 4-1.

Internal blast damage is proportional to the ratio of the high explosive to the volume of the space containing the burst. Therefore, it is possible to accomplish similar damage with a small charge weight in a small volume as with a large charge weight in a proportionately larger volume. It is to be noted that damage by internal detonations is caused not only by the air blast wave but also by the expanding gaseous products of detonation. Damage from the latter cause is also dependent upon the ratio of the high explosive released and the volume containing the burst.

A similar action occurs with an external blast except that the explosive action is directed against the outer rather than the inner surface of the structure. With external blast, the overpressure produces an implosive effect on the target structure rather than explosive, and structural failure is generally due to inward collapse. See Figure 4-1.

The peak overpressures and the positive impulse of air shock waves from the detonation of spherically shaped explosive charges of 50/50 Pentolite (see Appendix) have been measured under ambient atmospheric pressures and temperatures simulating altitudes from sea level to 50,000 feet. Reference 4-2.a The following equations fit the experimental data.

$$\frac{P}{P_o} = \frac{37.95}{ZP_o^{1/3}} + \frac{154.9}{(ZP_o^{1/3})^2} + \frac{2034}{(ZP_o^{1/3})^3} + \frac{403.9}{(ZP_o^{1/3})^4} \quad (4-2.1)$$

$$\log_{10} \frac{I}{P_o^{2/3} C^{1/3}} = 1.374 - 0.695 \log_{10} (ZP_o^{1/3}) \quad (4-2.2)$$

where:
- P = peak pressure in psi
- P_o = ambient pressure in atmospheres (1 atmosphere = 14.7 psi)
- $Z = R/c^{1/3}$ known as scaled distance
- R = distance from explosion in feet
- c = weight of Pentolite in pounds
- I = positive inpulse, milliseconds psi

These equations are graphically represented in Figures 4-2 and 4-3. Later data beyond the range covered by equation 4-2.2 and Reference 4-2.a cause extensions of the line (on log paper) to be curved.*

The foregoing data on peak pressure and impulse may be corrected for other explosives by using the following relative values on an equal volume basis. Reference 4-2.b.

To obtain relative values on an equal weight basis, the specific gravity of the explosives must be considered. For example, the following table illustrates the values of peak pressure and impulse relative to composition "B" on both a weight and volume basis for HBX-1 and H-6. The exact composition and properties of the various explosives may vary slightly between the armed services.

The steel case retaining the explosive charge reduces the effectiveness of the charge since it requires energy to rupture it after the charge has been detonated. The effect of this has been studied in Reference 4-2.c with the following empirical results. For peak pressure:

$$\frac{W'}{c} = 1.19 \left[\frac{1 + M_r (1 - M')}{1 + M_r} \right] \quad (4-2.3)$$

For positive impulse:

$$\frac{W'}{c} = \frac{1 + M_r (1 - M')}{1 + M_r} \quad (4-2.4)$$

where:
- W' = equivalent bare charge weight
- c = actual charge weight
- m = actual casing weight
- M_r = ratio of casing weight to charge weight in cylindrical section
- M' = casing-to-charge weight-ratio parameter, defined as follows:

$M' = 1.0$ when $M_r \geq 1.0$ $\quad M' = m/c$ when $M_r < 1.0$

The types of targets likely to be encountered in blast warhead applications vary from light structures such as aircraft and frame buildings to intermediate structures such as vehicles and masonry buildings through very heavy structures such as armored vehicles and reinforced concrete structures. Light structures are critically damaged by blast when detonated in the air nearby (by proximity fuzes)

---

* References are listed at the end of sections within chapters.

Table 4-1 Characteristics of Explosives

Explosive	Relative Peak Pressure	Relative Impulse	Specific Gravity
Pentolite (50/50: TNT/PETN	0.98	0.97	1.68
Composition "B" (60/40/1: RDX/TNT/ Wax added)	1.00	1.00	1.68
TNT	0.92	0.94	1.60
Tritonal (80/20: TNT/Al)	1.04	1.08	1.70
Torpex 2 (42/40/18: RDX/TNT/Al)	1.13	1.16	1.76
HBX - 1 (67/11/17/5/0.5: Comp B/TNT/Al powder/D-2 Desens/CaCl)	1.12	1.19	1.72
H-6 (74/21/5/0.5: Comp B/Al/D-2 Desens./CaCl)	1.20	1.39	1.75

Table 4-2 Characteristics of HBX-1 and H-6

Explosive	Relative peak pressure on basis of		Relative Impulse on basis of	
	Weight	Volume	Weight	Volume
HBX-1	1.07	1.12	1.14	1.20
H-6	1.12	1.19	1.30	1.39

or by surface impact close to the structures. In this instance, the warhead case, acting as a pure container, represents from 15% to 20% of the total warhead weight. If a warhead with such a case registers a direct hit on light target structure, the case is usually strong enough to properly retain the charge until detonation takes place and results in a damaging high order explosion. Warheads for use against intermediate structures are usually fuzed to explode on impact or soon thereafter. For such a warhead, the case is strengthened somewhat to withstand the impact forces and usually represents about 25% to 30% of the total warhead weight. Warheads used against very heavy structures are designed with heavy steel ogival heads and reinforced walls to give them penetration capability. For this use, the case may represent 50% of the total warhead weight.

The effects of altitude on external blast are reported in References 4-2.n and 4-2.o. The blast envelope generally takes the form of an oblate spheroid. The axial bounds of this envelope are generally unaffected by altitude; the transverse bounds (above and below the target) are generally pinched in with increase of altitude. More extreme effects (on the transverse bounds) are caused by gust loading due to the velocity of the target. Reference 4-2.o.

If the direction of the external blast relative to a target surface is face-on, the blast volume is much greater than if the direction is side-on, especially at high altitudes. In general, for bursts occurring at equal distances from an aerial target surface, the damage is a direct function of the charge weight. As this weight decreases from large (≈ 600 lb) to small (≈ 100 lb), the tendency is to obtain local failure instead of drastic and immediate disruption of the aircraft.

It is to be noted that many aircraft are capable of continuing in flight with considerable local damage. In utilizing the data presented in References 4-2.n and 4-2.o, it is to be further noted that the Russian IL-38 "Bear" and present Russian fighters are generally similar to the B-29 and F-86 aircraft, respectively, in regard to the effects of blast phenomena.

The effects of altitude on internal blast are reported in References 4-2.p and 4-2.q. These references report the results of experiments conducted by using small charge weights against various aircraft components, in which the blast waves struck the nearest portion of the structure normally (i.e. head on) but do not include the effect of charge velocity, i.e., far side enhancement. These results show that the ratio of the explosive weight needed at high altitude to that at sea level for equal damage increases with altitude. The average ratios for the aircraft components tested were 1.22, 1.39, and 1.72 for altitudes of 30,000, 45,000, and 60,000 feet respectively.

References 4-2.r through 4-2.t report the effects of the charge velocity on the resulting peak pressure and positive impulse. These experiments are all with 3/8 lb charges at sea level. They indicate that the side-on peak pressure and positive impulse are both increased in the direction of charge motion and decreased in the opposite direction relative to results obtained from detonation of a stationary charge. These velocity effects are probably larger for the relatively small charge weights tested than for the larger weights used in engagements. Similarly, the effects are likely to be larger at high altitudes than at sea level.

The effects of altitude on target surface bursts are intermediate between those for external and internal blast. The effect of a surface burst of a given charge weight may be approximated by the damage due to an internal burst of one-half this weight.

The damage to industrial buildings from external blast was studied experimentally in Reference 4-2.d. The damage from blast was negligible at overpressures up to 2.0 psi and consisted almost entirely of broken windows and roof decks. At 3.5 psi all windows and roof decks were broken and some walls cracked but did not cave in. At 5.0 psi a few localized portions of external walls were blown down. At 7.5 psi over half of the walls crumbled and parts of the roof structure including framing were brought down. At 10 psi all of the masonry walls were reduced to rubble and the steel support structure was distorted; only the major steel columns were left standing. At 15 psi the entire building had collapsed and everything was wrecked except equipment in the basement and some steam generators above ground. At 30 psi the entire building and everything above ground with the exception of the steam generators was a tangled mass of masonry and crumpled metal.

Data on the penetration ability of various ogives is given in Table 4-3. This information may be used as a guide to designing penetration cases.

The warhead designer, acting as a member of a weapon system team, is given information defining the targets, the missile performance, the guidance accuracy, and the allowable total warhead weight. With this known and by use of the foregoing data on blast effects and casings, a decision is made regarding the function of the warhead case between the limits of a pure container and a containing-plus-penetration means. This will establish the case configuration. Also, using the data and percentages given, an approximation of the weight of the case is made.

Compatibility of Weight and Space Allocated Knowing the approximate configuration and weight of the case and the total allowable warhead weight, an approximate charge-to-metal ratio may be established. Having this, the compatibility of the weight and space allocated to the warhead is checked using the following formula:

$$\frac{V}{W} = \frac{1}{\rho_m(1+c/m)} + \frac{c/m}{(1+c/m)\rho_c} \quad (4\text{-}2.5)$$

where:
- V = total warhead volume in cubic inches
- W = allowed weight of warhead in pounds
- c/m = charge-to-metal ratio based on total charge and metal weight
- ρ_m = density of metal in pounds per cubic inch
- ρ_c = density of charge in pounds per cubic inch

Use of the above equation will indicate whether weight or space is the limiting factor in determining the size of the warhead case. Weight will be the limiting factor when total weight of a warhead which occupies the entire warhead compartment is in excess of the weight allocated for the warhead. In this instance, the weight and size of the warhead must be reduced accordingly. Space is the limiting factor when the total weight of the warhead is less than that allocated for the warhead. In this instance the charge-to-metal ratio may be decreased for penetration type cases by increasing the amount of steel in the ogive. This will increase its penetration capabilities and increase the weight per unit volume of the warhead. Note that the method of mounting the warhead can also be a limiting factor in determining the weight and shape, and consequently the mounting must be considered. At this point in the design, the overall configuration of the warhead, total weight and charge-to-metal ratio may be fixed.

Installation and Handling Provisions The installation provisions in the warhead compartment and support fixtures required for handling must be studied. The requirements for these will be obtained from the missile system designer. Typical installation and support fittings for the warhead consist of mounting lugs or a mounting ring around the periphery of the warhead case. These mounting fixtures will normally be located centrally on the warhead case or near both ends of the case. The warhead may be supported from one end only, but additional strength in the casing is then necessary to overcome the cantilever effect of the overhanging portion.

Strength Analysis The strength of the overall case and the support fittings can now be analyzed. The type and magnitude of the loads to which the warhead will be subjected depends on the location of the support fittings and the design load criteria. This information is also obtained from the missile system designer. In some installations the warhead case is an integral part of the missile structure and must be treated accordingly in the analysis. Where the case functions only as a container for the explosive charge the design is based on a stress analysis considering the missile and handling inertia load factors. For such warheads, missile-acceleration forces are always important, and centrifugal forces cannot be neglected for spinning rockets and missiles. When impact or penetration of structure is required, it will usually be found that the impact or penetration loads are much more severe than the missile and handling loads. Under these conditions the impact and penetration loads determine the strength of the warhead case, while the missile inertia load factors determine the design of mounting lugs from a strength viewpoint.

Explosive Loading and Sealing An opening must be provided in the warhead casing to allow loading of the explosive charge. Loading apertures are usually centrally located on either end of the warhead case when target impact is not required. When target penetration is required, the loading apertures will normally be on the rear of the warhead case, because such an aperture in the nose weakens the penetration ability of the warhead. The inside surface of the warhead casing is coated with inert material to eliminate chemical reaction between the explosive and warhead metal, to provide

Table 4-3

Penetration Capabilities Of Penetration Case

Description of Projectile	Approx. Weight lb	Projectile Diameter in.	Impact Velocity fps	Impact Angle deg.	Material Penetrated	Penetrated Depth – in.	Reference
Bomb, Experimental, Mk 81, Mod O	250	9	1000	20	STS Armor*	1-7/8	9-2.i
Bomb, G.P. (Low Drag) Type EX-12, Mod O	500	10.95	1000	20	STS Armor*	1-1/4	9-2.j
Bomb, G.P. (Low Drag) Type EX-10, Mod 3	925	-	1000	0	STS Armor*	1-1/2	9-2.l
Bomb, G.P., AN-M65A1	1240	19	807	0	STS Armor*	2	9-2.k
Warhead, G.P., T23	1675	-	1487	0	STS Armor*	2	9-2.f
Warhead, G.P., T23	1675	-	1486	15	STS Armor*	1.5	9-2.f
Warhead, T.P., T23	1691	-	1392	45		1	9-2.f
Bomb, G.P. (Low Drag) Type EX-11, Mod O	2000	17.95	1000	20	STS Armor*	1-7/8	9-2.h
Bomb, G.P., AN-M66A2	2402	24	574	0	STS Armor*	2	9-2.k
Bomb, General Service, T55	3000	24	1000	15-30	STS Armor*	.633	9-2.g
Bomb, G.P. (Low Drag) Type EX-10, Mod 3	925	-	1000	0	Concrete**	24	9-2.l
Inert Loaded Warhead, T2	1400	-	1128	0-20	Concrete**	10	9-2.l
Bomb, General Service, T55	3000	24	1000	15-30	Concrete**	10	9-2.g
Bomb, T28E1 (Amazon II)	25000	38	1070	17	Concrete**	177	9-2.m
Bomb, T28E2	25000	32	1090	20	Concrete**	249	9-2.m

*STS Armor Plate
**Concrete, Reinforced

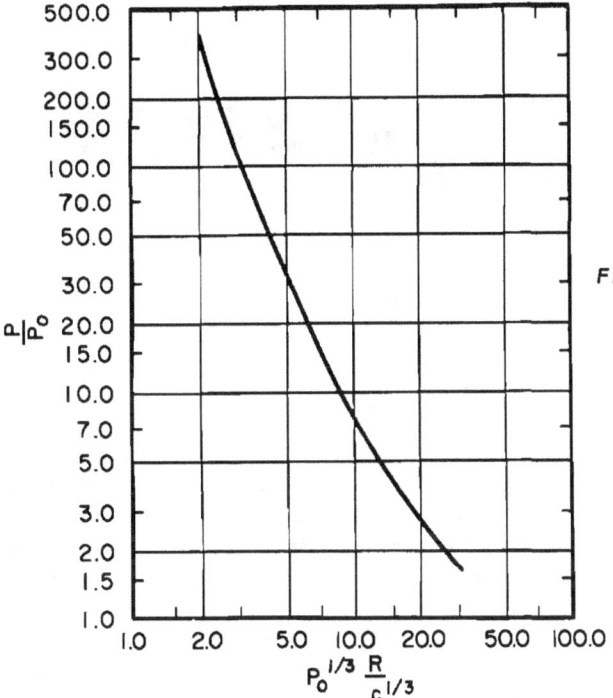

Figure 4-2. Peak Pressure Vs. Scaled Distance at Various Atmospheric Pressures, 50/50 Pentolite Spherical Bare Charges

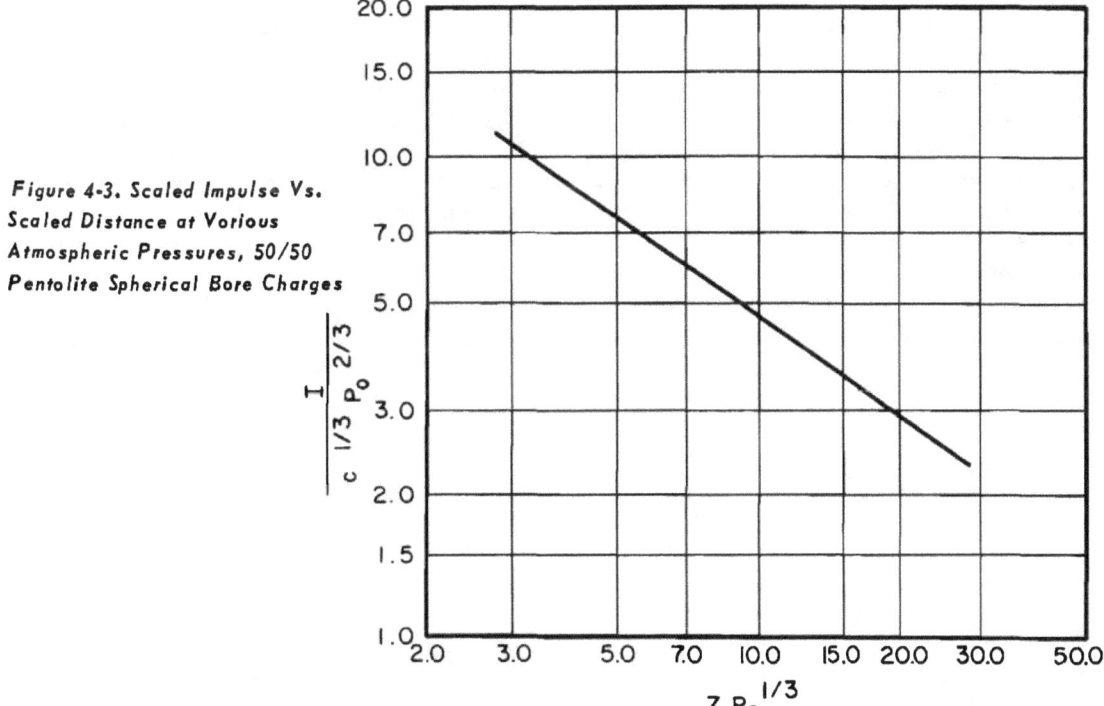

Figure 4-3. Scaled Impulse Vs. Scaled Distance at Various Atmospheric Pressures, 50/50 Pentolite Spherical Bare Charges

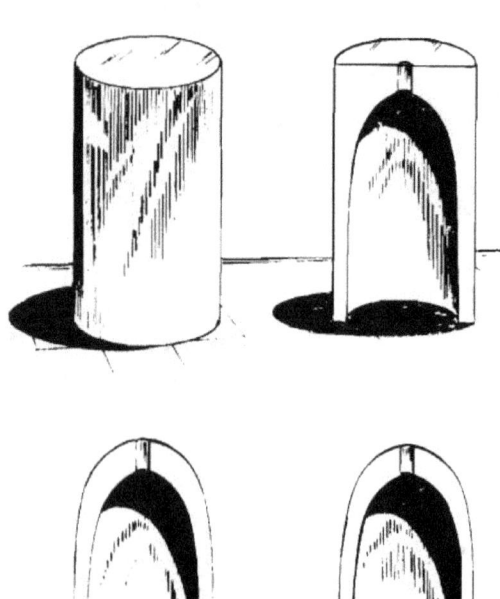

Figure 4-4. One Piece Fabrication

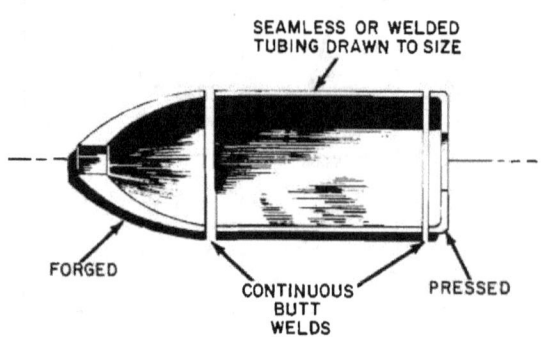

Figure 4-5. Multipiece Fabrication

a cushion between small crevices or projections of the metal surface, and to provide a bond between the explosive and the metal casing wall.

The explosive charge is usually cast in place in the warhead cavity. The explosive is heated until molten, poured in place, and then allowed to cool and solidify. The pellet-and-pour method of loading is frequently used in larger warheads weighing more than 200 pounds. This method involves alternately pouring the explosive melt at a temperature somewhat greater than the normal melt temperature, and dropping in quantities of the same explosive in pellet form until the warhead is filled. Loading may also be accomplished by pressing or tamping preformed or plasticized explosive in place.

Variations in density caused by porosity and shrinkage cavities are most undesirable. These tend to affect the design weight and also, more importantly, degrade the initial fragment velocity. Filling is generally accomplished near the freezing temperature of the explosive to avoid excessive shrinkage and to control the grain size which affects sensitivity. Shrinkage is minimized by the use of risers, controlled cooling rate, etc. The final seal usually consists of a plate bolted in place over the loading aperture. A layer of inert material is usually inserted between the sealing plate and the explosive charge to allow for thermal expansion.

Fuze Installation The warhead fuze is usually located in either the nose or the rear portion of the warhead. Some warhead applications require a fuze in both locations. A threaded hole for fuze insertion can be tapped directly in the warhead casing. A fuze adapter consisting of a bushing with external and internal threads is generally used. The fuze is threaded into the adapter, or an adapter plug is inserted to keep out foreign material and moisture prior to fuze installation. This plug can also be so made as to be useful in handling the warhead.

Fabrication and Tooling Blast warheads are fabricated using two general methods--one piece and multipiece construction.

One piece construction utilizes the so-called pierce and draw method. A preheated billet is pierced with a mandrel sufficiently to start the general internal shaping. The front ogive is then formed. The billet is then forced through draw rings to form the cylindrical and rear portions. This is followed by heat-treating until minimum physical properties are met. See Figure 4-4.

For the multipiece construction, the nose and rear sections may be either pressed from plate or forged, depending on the configuration. The cylindrical section is usually fabricated from seamless or welded tubing. The above components are then assembled by butt welding. See Figure 4-5.

Fabrication by the pierce and draw method is best adapted to heavy-walled munitions while thin-walled cases are usually fabricated using multipiece construction. After initial setup, higher production rates can be obtained with the pierce and draw procedure.

Summary of Fuzing Requirements The fuze designer needs design information to design a fuze which is compatible with the missile system and the warhead. He will have access to the same missile system data as did the warhead designer. In addition to this, the fuze designer will need the following information relating specifically to the warhead:
(1) Weight of warhead allotted to fuze or S&A
(2) Type of explosive used
(3) A drawing of the warhead
(4) Type of case used, that is pure container, impact, or penetration
(5) V_m, V_t

Summary of Design Data At the conclusion of the design procedure, a summary of engineering data relating to the warhead should be prepared. This should include the following items:
(1) Total weight
(2) Design and installation drawings
(3) Explosive
  (a) Material
  (b) Weight
  (c) Density
(4) Charge-to-metal ratio
(5) Case
  (a) Type
  (b) Weight
(6) Location of center of gravity
(7) Mounting means

4-2.3. References

4-2.a "The Effect of Atmospheric Pressure and Temperature on Air Shock", Jane Dewey and Joseph Sperrazza, BRL Report 721, May 1950.

4-2.b "Relative Air Blast Damage Effectiveness of Various Explosives", W. E. Baker and O. T. Johnson, BRL Report 689, June 1953.

4-2.c "The Effect of the Steel Case on the Air Blast from High Explosives", E. M. Fisher and C. J. Aronson, NAVORD Report 2753, Feb. 1953.

4-2.d Report by Armour Research Foundation of Illinois Institute of Technology, "Study of Vulnerability of a Thermal Electric Power Plant to Air Blast", Mar. 1954.

4-2.e "Final Report on Ballistics Experimental Tests of Guided Missile Warheads", NPG Report no. 691, Dec. 1950

4-2.f "Tests of Warheads, G. P., 1500 lb, T23", NPG Report no. 1234.

4-2.g "First Partial Report of Gun Firing Test of Bombs, General Service, 3000 lb, T-55", NPG Report no. 945, Mar. 1952.

4-2.h "Thirty second Partial Report on Bombs and Associated Components. Final Report on Plate Penetration Tests of the 2000 lb, G. P. (Low Drag) Type EX-11, Mod O", NPG Report no. 950.

4-2.i "Plate Penetration Tests of Experimental 250 lb, Mk 81, Mod O Bomb Bodies", NPG Report no. 1292, Sept. 1954.

4-2.j "Plate Impact Tests of Low Drag, G. P., 500 lb, Bomb Type EX-12, Mod O", NPG Report no. 1299, Oct. 1954.

4-2.k "Ballistic Test of Modified 1000 lb, G. P. Bomb, AN-M65A1 and 2000 lb, G. P., Bomb, AN-M66A2 and Bomb Data for Penetration of Missile Resistant Armor", NPG Report no. 382,

4-2.l "Ballistic Tests of 1000 lb Low Drag, G. P. Bomb, Type EX-10, Mod 3", NPG Report no. 748, April, 1951.

4-2.m "Penetration and Deceleration of 25,000 lb Bombs in Massive Concrete Targets", BRL Report 712, Dec. 1949.

4-2.n "A Method of Predicting External Blast Vulnerability of Aircraft as a Function of Altitude with Application to B-29 Aircraft", O. T. Johnson, et al., BRL Report 1002, Dec. 1956.

4-2.o "An Experimental Investigation of the Effect of Motion of a B-29 Horizontal Stabilizer on External Blast Damage from Explosive Charges", R. L. Ballard, et al., BRL Report 982, June, 1956.

4-2.p "Internal Blast Damage to Aircraft at High Altitude", J. Sperrazza, BRL Memo. Report 605, April, 1952

4-2.q "Internal Blast Damage to Aircraft at High Altitudes--Part II", W. E. Baker and L. E. Needles, BRL Memo. Report 1036, August, 1956.

4-2.r "Air Blast Measurements Around Moving Explosive Charges", J. D. Patterson II and J. Wenig, BRL Memo. Report 767, March, 1954.

4-2.s "Air Blast Measurements Around Moving Explosive Charges, Part II", B. F. Armendt, Jr., BRL Memo. Report 900, May, 1955.

4-2.t "Air Blast Measurements Around Moving Explosive Charges, Part III", B. F. Armendt and J. Sperrazza, BRL Memo. Report 1019, July, 1956.

4-2.4. Bibliography

(1) "Report on Tests of the Effect of Blast from Bare and Cased Charges on Aircraft", James N. Sarmousakis, BRL Memo. Report 436, July, 1946.

(2) "Internal Blast Damage to Aircraft at High Altitude", J. Sperrazza, BRL Memo. Report 605, April, 1952.

(3) "Studies of the Influence of Variations of Blast and Structural Parameters on Blast Damage to Structures", E. Sevin and R. W. Sauer, Armour Research Foundation, ASTIA AD-47 930, Sept. 1954.

(4) "Air Blast Loading on Structures", D. C. Sachs and S. R. Hornig, Stanford Research Inst., ASTIA AD-43 119, July, 1954.

(5) "A Simple Method for Evaluating Blast Effects on Buildings", Armour Research Foundation, ASTIA AD-38 891, July, 1954.

(6) "Deformation Model Studies of Buildings Subjected to Blast", H. Williams and A. Ruby, Vibration Research Lab., ASTIA AD-58 619, Dec. 1954.

(7) "Effects of Impacts and Explosions", NDRC-DIV. 2 Vol. 1.

4-3. FRAGMENTATION WARHEADS

4-3.1. Detail Design Steps

Step No. Detail Design Step
1. Estimate the Optimum Beam Width
2. Select the External Configuration
3. Compute the Maximum Allowable Charge to Metal Ratio
4. Compute the Maximum Possible Fragment Ejection Velocity
5. Select the Optimum Fragment Mass and Ejection Velocity
6. Compute the Actual Charge to Metal Ratio and Select the Explosive Type
7. Select the Fragment Shape and Material
8. Select the Method of Fragment Control
9. Design in Detail the Fragmenting Metal
10. Design in Detail All Other Components
11. Prepare Summary of Fuzing Requirements
12. Prepare Summary of Design Data

The exact order of the design procedure may vary depending upon the viewpoint of the designer and, even more, on the military requirements which often fix certain parameters in advance.

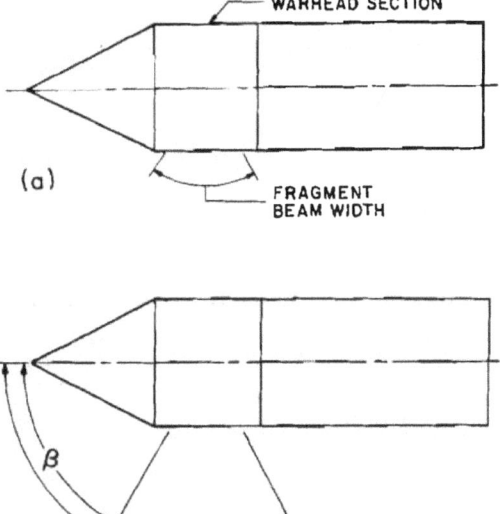

Figure 4-6. Definition of Fragment Beam Width

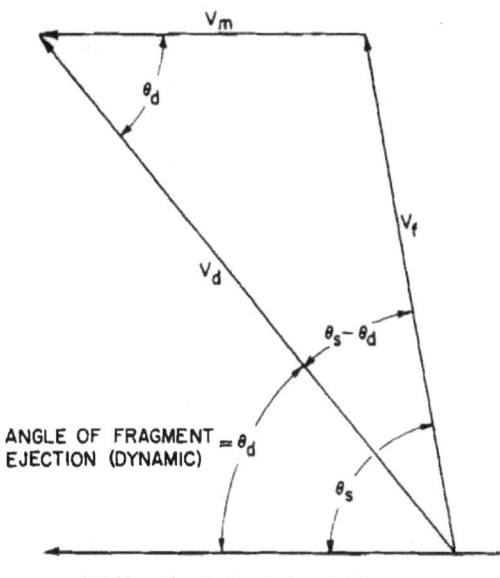

Figure 4-7. Vector Addition of Fragment and Missile Velocities

4-3.2. Detail Design Data (Fig 4-6 through Fig. 4-13; Tables 4-4 through 4-7)

Beam Width In designing a fragmentation warhead, the designer may first estimate the required fragment beam width. Fragment beam width is defined as the angle covered by a useful density of fragments, as shown in Figure 4-6(a). It also is sometimes given as shown in Figure 4-6(b), in which case the actual beam width is $\alpha - \beta$

To better understand the behavior of the fragments, it is necessary to know what happens to them between the time of detonation and the time of their arrival at the target. An analysis of the dynamic fragment velocity and the method for the determination of the fragment pattern follows.

Upon detonation of the explosive charge, the detonation wave causes the explosive and its case to swell until the failure point is reached. The case then fails in shear and tension and fragments are ejected at high velocity. If the warhead is stationary at the time of detonation, the velocity possessed by the fragments after a short travel is termed the "initial static fragment velocity". The method of computing this velocity will be shown later. If the warhead is moving through space at the time of detonation, as is naturally the case in flight, the velocity possessed by the fragments a short distance away is termed the "initial dynamic fragment velocity". This dynamic velocity is obtained by adding vectorally the static fragment velocity and the missile velocity.

Initial Dynamic Fragment Velocity The initial dynamic fragment velocity, V_d, is found by applying the law of cosines in Figure 4-7 and is given by

$$V_d^2 = V_f^2 + V_m^2 + 2 V_f V_m \cos \theta_s \quad (4-3.1)$$

where:

V_f = static velocity of the fragments
V_m = missile velocity
θ_s = angle of fragment ejection (static), to be derived later

CLASSIFICATION		MACH NUMBER RANGE							
		0 to 0.6		0.6 to 1.4		1.4 to 4.4		4.4 & Higher	
		C_D	C	C_D	C	C_D	C	C_D	C
Balls		.245	.00093	.41	.00155	.48	.00182	.456	.00172
Rt. Cyl. & Cubes	S=1	.330	.00155	.50	.00235	.57	.00267	.530	.00249
Long Fragments	S=5	.330	.00194	.50	.00294	.57	.00335	.530	.00312
	S=10	.330	.00233	.50	.00354	.57	.00403	.530	.00375
	S=15	.330	.00263	.50	.00399	.57	.00455	.530	.00423
	S=20	.330	.00287	.50	.00435	.57	.00496	.530	.00461

Table 4-4 One Piece Fabrication

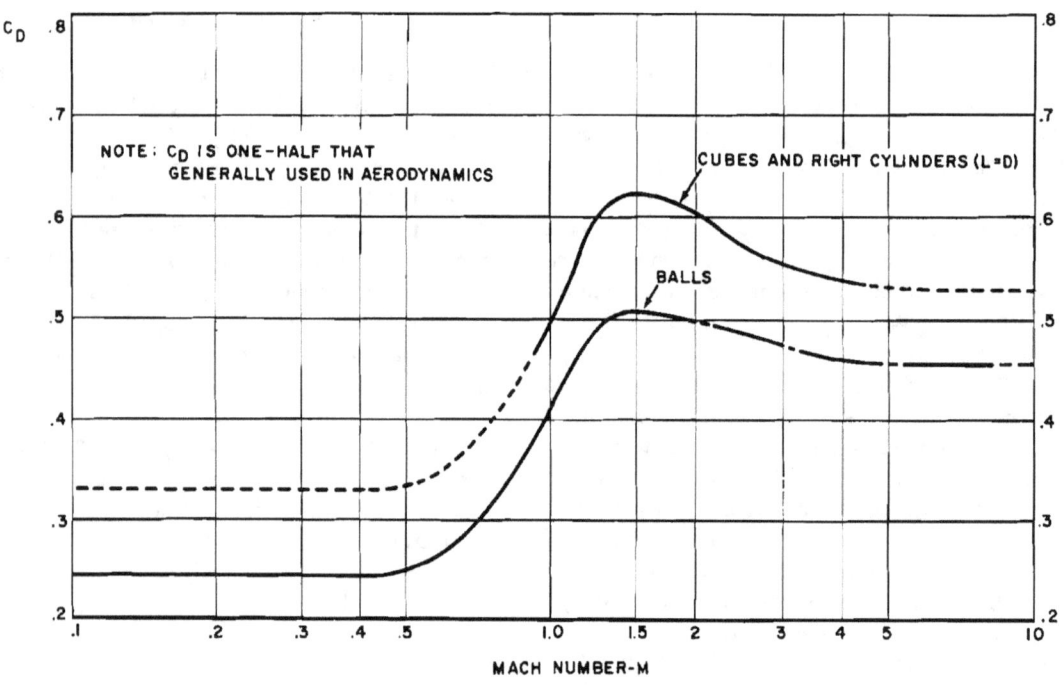

Figure 4-8. C_D Vs. Mach Number for Various Fragment Types

As a fragment travels through the air, it is slowed down by air resistance so that it will strike a stationary target at a lower velocity in free air than the initial velocity. For moving targets, the striking velocity is actually the vector difference of the target velocity and the fragment velocity at the end of its travel, and hence the striking velocity may be greater than the initial fragment velocity. This is discussed further in Section 5-2.3. The actual trajectory of the fragment in space can generally be ignored since its path is so short that the effect of the gravity can be neglected, and a straight line space trajectory within the fragment's lethal distance can be assumed.

Fragment Slow-down If the initial fragment velocity in free air is known, the velocity corresponding to a given distance traveled can be computed as follows:

$$V_x = V_o e^{-C_D \rho_a R_D (A/M_f)} \qquad (4-3.2)$$

where, in consistent units:
- R_D = distance traveled
- V_x = velocity at any distance R_D
- V_o = initial relative velocity in free air
- A = random projected area of fragment
- ρ_a = air density
- M_f = fragment mass
- D = drag force
- C_D = drag coefficient in dimensionless units = $D/\rho_a A V^2$.

C_D is one-half the drag coefficient generally used in aerodynamics. Equation 4-3.2 involves the assumption that C_D is substantially constant over the travel distance R_D.

A curve of C_D vs Mach number is given in Figure 4-8. To obtain the distance R_D for given initial and final velocities, the value of C_D corresponding to the mean velocity, $(V_x + V_o)/2$ is read from the curve. The value of R_D is then obtained from equation (4-3.2). If the variation in C_D is relatively large, the velocity range should be divided into subranges (such as subsonic, transonic, and supersonic) for which C_D is approximately constant and the distances added to obtain R_D.

In many instances the distance R_D will be given and the velocity (V_x) at this distance will be sought. In using equation 4-3.2 it is necessary to use a C_D corresponding to an average velocity over the distance, which will require a few iterations of a trial and error method of solution. However, the velocity (V_x) is not very sensitive to C_D when the distance R_D is small.

Also of some interest is the average velocity $(\bar{V})$ over the distance R_D:

$$\bar{V} = \frac{V_o K R_D}{e^{K R_D} - 1}$$

where:
$$K = C_D \rho_a A/M_f$$

This equation is most often used in finding the values of K or C_D from tests that provide values of V_o and $\bar{V}$ over a measured R_D. It can be reversed to give the initial velocity if the average velocity and drag coefficient are known.

In connection with Figure 4-8, the following conventions are used:

(a) For balls or spheres, the projected area is that of the maximum section, or $A = \pi r_o^2$ where r_o is the radius in feet.

(b) For cubes (and also approximately for right cylinders where length $L = D$) the curve is used in conjunction with an area given by $A = 0.25$ times the fragment total area in square feet.

(c) Tests at Mach 5.8 indicate that there is no significant difference above Mach one between the drag coefficient (C_D) of cubes and elongated fragments approximating rectangular parallelepipeds having a length of approximately 9.5 times the geometric mean of width and thickness. Hence to find C_D for an elongated fragment, one should use for supersonic velocities, either the drag coefficient for a cube with an area $A = 0.25$ times the fragment total area in square feet or a closely consistent relation such as

$$V_x = V_o e^{-0.0045 (\rho_a/\rho_o) R_D / \sqrt[3]{M_f}} \qquad (4-3.3)$$

where (ρ_a/ρ_o) is the air density ratio and M_f is the fragment mass in ounces.

For steel fragments (spheres, cubes, or rectangular parallelepipeds) the following simplified equation is more convenient to use than equation (4-3.2):

$$V_x = V_o \, e^{-C(\rho_a/\rho_o)R_D/\sqrt{M_f}} \qquad (4-3.4)$$

where M_f is given in ounces and R_D in feet. For various Mach number ranges, Table 4-4 gives values of C_D and C for fragments classified according to shape and a parameter S, where S is the ratio of length to geometric mean lateral dimension.

Values of C_D for irregularly shaped fragments are not accurately known but are possibly slightly higher than those of the oblong, square-cornered shapes considered above. Drag coefficients for fin stabilized fragments of unusual shape should be obtained from aerodynamic analysis or tests; however for darts similar to those of around 8-gr. designed by the International Harvester Company (IHC Report 15), values of C_D can be obtained from References 4-3.z and 4-3.aa.

Fragment Patterns Both for prediction and design of fragmentation warheads, it is imperative to know how the pattern of fragments ejected from the warhead is related to the design of the warhead. The primary dependence is on the shape of the warhead wall and the location of the point where detonation is initiated.

Except perhaps for the detonation point, fragmentation warheads are nearly always symmetrical about a longitudinal axis, which is usually also the axis of the missile carrying the warhead. Correspondingly, it is usually assumed that the fragmentation pattern is symmetrical about the same axis. In the case of truly symmetric warheads the available evidence does not contradict the hypothesis of symmetry of the fragment pattern, though there is only a little experimental evidence on this point, most effort having been concentrated on determination of the variation in the other direction as discussed in the next paragraph.

In cases of asymmetric staggering of notches in the casing, casings made in more than one part, or asymmetric location of the point where the detonation is initiated, there are some indications of asymmetry in the fragment pattern. However, the only case in which the problem appears serious is that of a very asymmetric detonation point, especially if the warhead is annular in shape (i.e., has a large hollow space along the axis). In this case, the detonation wave may strike the casing at substantially different angles on the near and far sides, producing correspondingly different patterns; moreover, in the zone where detonation waves traveling around opposite sides of the annulus meet, fragment shatter and alteration of velocities are to be expected.

Reverting to the usual case of axial symmetry, it remains to consider the fragment density as a function of angle of emission ϕ measured from the forward direction of the warhead axis. Of interest are two different versions of this pattern, usually called "static" and "dynamic". The static pattern is the one produced if the warhead is detonated while motionless, while the dynamic pattern is the one obtained if the warhead is in flight.

For the prediction of static fragment patterns, reliance is customarily placed on the Shapiro method, Reference 4-3.c. This method assumes that fragments are (or can be thought of as) originally arranged in successive rings, the part of the warhead casing of interest being composed of many such rings stacked one on another, each with its center on the axis of symmetry. Although this may not be the actual mode of fabrication of the casing, the Shapiro relation is probably a sufficiently accurate approximation for initial design purposes. Figure 4-9 shows a longitudinal cross-section of such a warhead; Figure 4-10 is a more detailed view of the cross-section in the vicinity of one ring, with pertinent variables labeled. Shapiro considers that the final static pattern is obtained by compounding a nominal angle of ejection with a dispersion about this nominal angle.

The fragments from a given ring are nominally ejected in a direction making an angle

ϕ with the forward missile axis where, theoretically, ϕ is given exactly by equation 4-3.6 and approximately by equation 4-3.7. The notation is the same as that shown in Figure 4-10, except that V_o is the initial fragment velocity (ejection velocity) and V_D is the velocity of the detonation wave in the explosive.

$$\frac{\phi}{2} = 90° - \arcsin\left(\frac{2aV_D + (c-d)V_o}{2bV_D}\right) \quad (4-3.6)$$

$$\frac{\phi}{2} \approx 90° - A - \arcsin\left(\frac{V_o \cos \beta}{2bV_D}\right) \quad (4-3.7)$$

Although the derivation of these equations is not entirely empirical, their results appear to be consistent with experiments.

The fragments nominally ejected at angle ϕ are subject to dispersion about this direction. The actual spread increases with the length of the warhead and the ejection velocity decreases at the ends. The standard deviation of the dispersion assumed for missile warheads is 3°. Figure 4-11 shows this distribution for a warhead having 10° beam width.

Having found the nominal direction of ejection of the fragments from each individual ring, and their dispersion about that ring, it remains to combine these to get the static pattern from the warhead as a whole. To determine the total number of fragments at a given static angle θ_s, measured as usual from the forward axis, and generally being the same as ϕ, the contributions from the various rings are added together. Reverting to Figure 4-7, $\angle \theta_s$ means the $\angle \phi$ of Figure 4-10. Actually one does not deal with the exact angle θ_s, but with an angular interval (such as ±1° increments in ϕ), represented by this angle. Having thus obtained the total number n_θ of fragments ejected in the interval, conversion to density $D(\theta_s)$ of fragments per unit solid angle (steradian) is made by the following equation which treats the density at the center of the 2° interval as representing the average density in the interval.

$$D(\theta_s) = 4.560 \, (\csc \theta_s) \, n_\theta \quad (4-3.8)$$

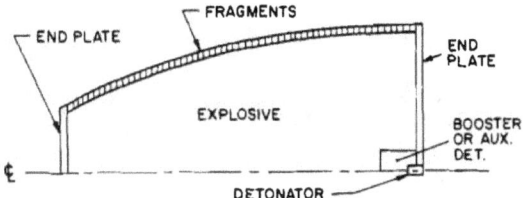

Figure 4-9. Longitudinal Section of a Typical Fragmentation Warhead

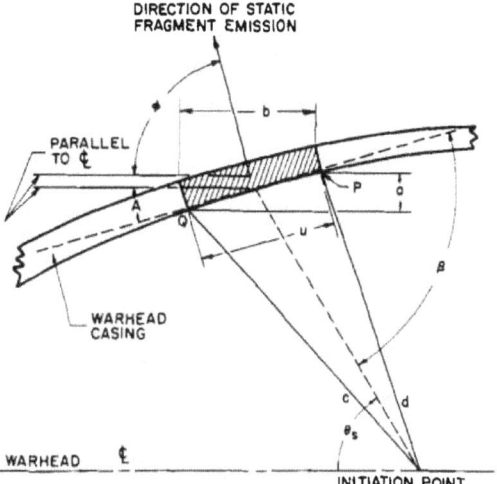

Figure 4-10. Diagram for Derivation of Angle of Emission of Fragments

(10° CYLINDRICAL WARHEAD, BASE INITIATED)

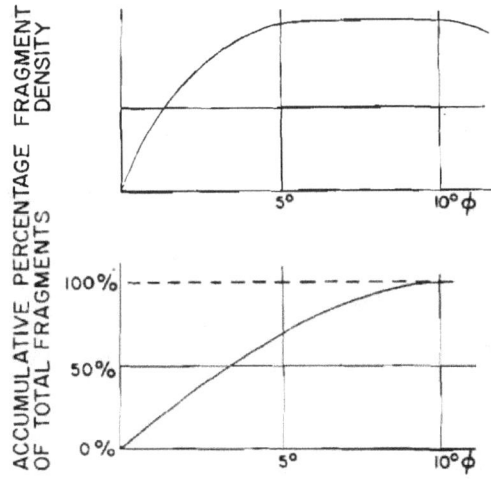

Figure 4-11. Distribution of Fragments about Nominal Ejection Direction

Having found the static fragment pattern, it now remains to find the fragment density if the warhead is moving through space and the vector velocities of the fragment (static) and warhead are to be compounded to find the actual direction in which the fragment proceeds outward. Reference 4-3.b.

The dynamic density $D(\theta_d)$ for a given direction θ_d is obtainable from the static density $D(\theta_s)$ for the corresponding direction θ_d Reference Figure 4-7 by the equations

$$\cot \theta_d = \cot \theta_s + \frac{V_m}{V_f} \csc \theta_s \qquad (4\text{-}3.9)$$

$$D(\theta_d) = D(\theta_s) \cdot \left(\frac{\sin \theta_s}{\sin \theta_d} \right)^3 \cdot \frac{1}{1 + \frac{V_m}{V_f} \cos \theta_s} \qquad (4\text{-}3.10)$$

where:
  V_f = velocity of fragments in the direction θ_s in the static case

and:
  V_m = warhead velocity or relative velocity of missile and target

In addition to the general characteristics of the fragment pattern, there is the question of the fine structure of the pattern: within small sections of the fragment beam, is the distribution of the fragments random, is it regular, or do the fragments tend to bunch? Generally speaking, a random pattern may be assumed, although in extreme cases this may give a different warhead effectiveness than a regular pattern. Bunching has sometimes been reported, but it seems likely that most cases of this have actually represented poor fragment control.

Consideration has sometimes been given to a "sweeping-up" effect as a result of target motion. That is, if the fragments are dispersed either laterally or along their trajectory (e.g., by velocity spread) then the motion of the target through the swarm may result in more hits on the target than if the target were motionless. It is believed, however, that this effect is negligible for missile warheads used against targets of relatively slow velocity, compared to other approximations usually involved.

In the event that the designer desires to calculate dynamic fragment densities using the relative velocity of missile and target rather than that of the missile alone, it is expedient to consider the component of target motion parallel to the missile motion (otherwise the pattern would be asymmetric). The relative velocity V_r is then often approximated by Reference 4-3.b

$$V_r = V_m - V_t \cos \theta \qquad (4\text{-}3.11)$$

where:
  θ = the angle between missile and target courses

and:
  V_t = velocity of target

For a more exact treatment of the relative velocity V_r, see Reference 4-3.bb.

Selection of Beam Width The factors influencing the choice of beam width are the target, standard error of guidance, aspect, and the fuzing accuracy. The first consideration should be given to the target. The information given the warhead designer will include the vulnerable area of the target which must be covered, or in the case of multiple vulnerable areas, the distance between vulnerable components. There may or may not be information on the fuzing accuracy. If not, one must specify the amount of dispersion along the trajectory that can be tolerated. The best fuzing accuracy understood to have been attained in tests to date was a standard deviation of approximately 15 feet. If no other data on the fuzing is available, a conservative distance of 25 feet can be used for a reasonable estimate against aerial targets. The beam width may also be affected by fuze location. Some safety and arming fuzes are side-mounted, in which case they interfere somewhat with the symmetry of the beam.

Knowing the target characteristics, error of guidance, and aspect, the designer may determine the necessary beam width graphically as shown in Figure 4-12. The design burst point should be designated as the mid-point of the target vulnerable length, with the appropriate allowance made for the guidance error. The beam width θ_b selected should contain 85 to 95 percent of the fragments, and cover the target projected vulnerable area for design burst points at distance σ_G from the target. If the fuzing accuracy is known, it should be incorporated as shown in order to estimate the approximate width of the fragment beam that will cover the target in the event of early or late detonation. In the event that the target is completely missed, or only a small portion of the beam covers the target when the fuzing accuracy is considered, the beam width should be enlarged slightly to give a reasonable target coverage (i.e., by 50 to 60 percent of the beam width) for the bounds of fuze-initiated bursts

In the case of ground targets, such as personnel, the beam width is usually selected as the maximum attainable so as to cover the greatest ground area with the largest number of lethal fragments. (This may require a nose-spray warhead, see Figures 4-35 et seq. and context.) The area covered is a function of aspect, burst height, and missile and fragment velocities. In some cases the design beam width may be based on a requirement for uniform fragment distribution in the target area, see References 4-3.n and 4-3.o.

External Configuration Since it is essential for the designer to have at least a rough idea of the warhead shape needed for the beam width he desires, some general comparisons of shape and beam width follow. The testing of actual warheads to date show results which correlate with intuitive reasoning. That is to say, a spherical shaped warhead will produce the widest beam, while a short cylindrical warhead with concave sides gives a focusing effect and a very narrow beam. Variations between these two extremes give beam angles roughly proportional to their variation, considering similar detonation points. It should be mentioned that, for a vertical axis, a surface for constant fragment density on the ground known as the Kent-Hitchcock Contour (Reference 4-3.o) has been developed for bombs, but has not been practically applied since standard geometric shapes lend themselves more readily to present manufacturing techniques (Reference 4-3.n) and are better suited to the smaller inclinations typical of missiles.

The majority of conventional anti-personnel warheads developed to date have been located in the nose of the missile and are spheroidal in shape, as diagrammatically shown in Figure 4-13(f). This allows for the greatest possible beam width or ground coverage. The Kent-Hitchcock Contour also shows promise for the special application of vertical fall, but as previously stated has not yet been applied. If the warhead is located in a section other than the missile nose, the modified barrel type shown in Figure 4-13(e) is recommended. The distribution of fragments is dependent to some extent on the position of the booster in the warhead.

The largest proportion of fragment warheads designed to date for use against aerial targets are barrel-shaped. This has resulted from both the fact that the designer is usually allotted a cylindrically shaped section located in the body of the missile and the fact that a desirably large beam width results from this shape.

It is generally desired that the explosive charge of a fragmentation warhead be solid rather than hollow. There is no serious objection to a small conduit down the center but a large hole leads to reduced fragment velocity, other things being equal. On the other hand, too small a diameter warhead may result in failure to develop the full power of the explosion; the minimum satisfactory outside diameter is approximately 2 to 5 inches. Likewise, the length should not be too small--in general, length-diameter ratios of less than 1.25 seriously reduce the average fragment velocity. However, some compromise in this respect is generally required in the interest of other needs of the missile design with little if any gain realized by increasing the ratio over 2.5.

In general, it is not desirable that the warhead be cylindrical in shape because this gives an excessively narrow fragment beam. If an ogival shaped section of the missile is allotted to the warhead, this may give a sufficiently large beam width for a matching exterior surface of the warhead; otherwise, it will probably be necessary to shape the warhead like a barrel, and cover it with a fairing, with some, though not great, waste of either weight or of fragment velocity. It is also desirable to avoid packaging anything massive outside the warhead, as the fragments ejected would lose too much velocity in passing through such external material. However, it is often necessary to provide electrical cabling past the warhead, either externally or through a central conduit, and sometimes fuze antennas must be located on the outer surface.

It is usually easier to attain a desired fragment pattern if the initial detonation point is somewhere near the center of the warhead. The main purpose of shifting the initiation point from the warhead center is to throw the center of the beam forward or aft as required.

Aspect	= side
L	= target projected vulnerable length ($\perp$ to the plane containing missile and target), feet
σ_G	= guidance error, feet
S	= design burst point
S_1 & S_2	= possible burst points, due to fuzing error
θ_B	= beam width, degrees
Z_1 & Z_2	= possible fuzing error, feet

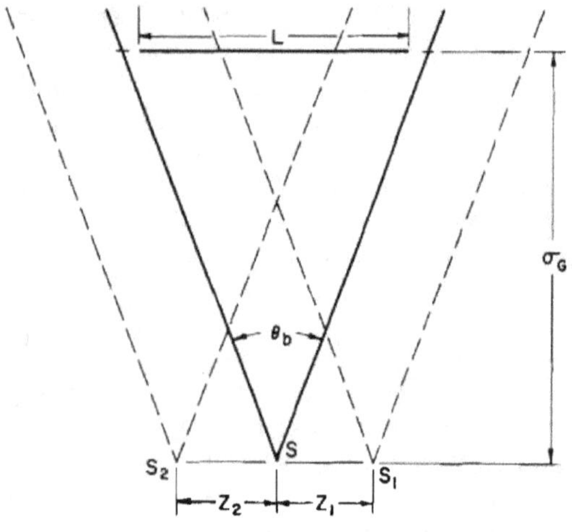

A series of examples of previous warhead designs are presented in Figure 4-13 in order to facilitate the selection of the proper warhead shape to obtain a specified beam width. These examples will guide one in selecting the approximate beam width; an exact design can only be determined after extensive calculations, and still must be proven by testing. For most cases of warhead initial design, this approximation should be adequate. However in each case, such design should be either verified or modified after testing.

Maximum Charge to Metal Ratio The charge to metal ratio, commonly referred to as c/m, is the ratio of the weight "c" of the explosive to

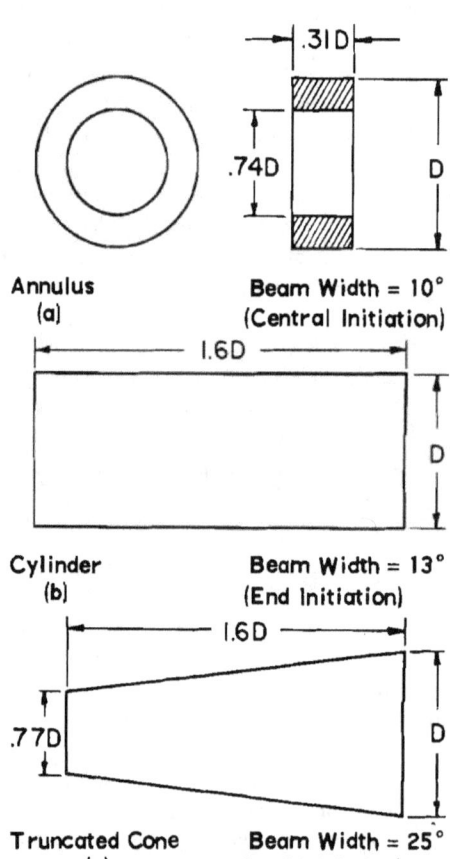

Figure 4-12. Graphical Solution of Optimum Beam Width

Figure 4-13a. Examples of the Effect of Warhead Shape on Fragment-Beam Width

the weight "m" of the metal case (excluding end plates, fittings, etc.). It will later be shown that the initial speed of the fragments emitted from the warhead is directly dependent upon c/m, subject to the warhead shape and the type of explosive employed.

Since the density of high explosive is approximately 22 percent of that of steel, the volume of a given weight warhead will vary as the charge to metal ratio is changed. The procedure for computing the maximum c/m which can be utilized in the weight and volume allotted for the warhead follows.

An approximate allowance must first be made for the so-called "dead weight" of the warhead, which is composed of non-fragmenting items such as the end plates, attaching fittings, required structure, detonator, etc. In warheads under 100 pounds built to date, this weight has varied considerably, from 10 percent to 32 percent in extreme cases where the warhead was required to carry large structural loads. It appears that a conservative estimate of "dead weight" for most warheads in the 100 pound class is 25 percent of the allowed gross weight. For warheads in the 100 to 300 pound class, this percentage may be lowered to between 10 to 20 percent. For warheads over 300 pounds an allowance of 10 percent should be reasonable. After the so-called "dead weight" components have been designed in detail, this weight estimate should be checked.

An approximation must also be made for the volume occupied by the dead weight. To compute this volume, an overall length allowance of 1-1/2 inches should be made for attaching fittings, end plates, etc. If a center tube is required through the warhead for missile wiring, it should also be considered as a "dead" volume. The usual diameter of such a center tube is 1.0 to 2.0 inches.

The useful or net weight (charge plus fragmenting metal) of the warhead may now be easily computed as follows:

$$W_n = W - W_{D.W.} \qquad (4\text{-}3.12)$$

where:
W_n = net weight
W = allotted total weight
$W_{D.W.}$ = estimated "dead" weight

Having previously established the shape of the warhead, and knowing the allotted warhead compartment dimensions, the total volume of the warhead may be computed. The useful or net volume (charge plus fragmenting metal) is given by

$$V_n = V - V_{D.W.} \qquad (4\text{-}3.13)$$

where:
V_n = net volume
V = total warhead volume
$V_{D.W.}$ = estimated "dead" volume

Once the net volume and net weight have been calculated, one may obtain the net specific volume (v) of the warhead, which is

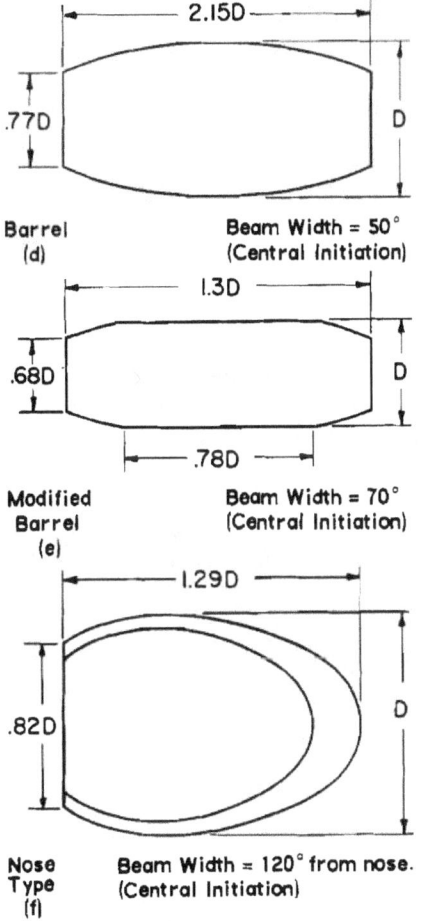

Figure 4-13b. *Examples of the Effect of Warhead Shape on Fragment Beam Width*

$$\nu = \frac{V_n}{W_n} \qquad (4\text{-}3.14)$$

Figure 4-14 illustrates the variation of c/m with warhead net specific volume. The maximum c/m that the allotted space and weight will permit may be read from this curve for the net specific volume just computed. The usual value for c/m is in the range of 0.2 to 0.5 for use against personnel, and from 0.4 to 2.0 for use against aircraft. Values of c/m as low as 0.1 may occur with small gun-boosted rockets used against ground personnel.

In succeeding sections of this handbook the optimum c/m from the viewpoint of fragment velocity and weight will be selected. If the optimum c/m is well below that allowed by the allotted space and weight, either the allotted volume is more than necessary, in which case "dead" space would be incorporated in the final design, or the warhead allowed weight is too low. Obviously either of these conditions are of interest to the missile system designer. It is to be noted that, many times, warheads are designed for missiles already in use, in which case the design would probably be carried out using the original warhead weight and volume. (In some cases, it is necessary to add ballast to bring the center of gravity to a position that stabilizes the missile.)

If the optimum c/m is above that allowed by the allotted space and weight, the converse would be true: either the warhead weight should be reduced or the volume increased. Reduction in warhead weight would mean fewer fragments and, hence, a lower warhead effectiveness. Another alternative is to design the warhead for other than optimum c/m. This alternative is usually acceptable because the effectiveness of a warhead is not highly sensitive to variations of c/m near the optimum.

Maximum Initial Static Fragment Velocity After the maximum allowable charge-to-metal ratio and the shape of the warhead have been established, the maximum initial fragment velocity may be estimated, that is, the velocity possessed by the fragments after they have been accelerated by the explosion. This occurs within a very short distance.

Four principal formulas are in use for predicting initial fragment velocities; Gurney's two formulas, developed for an infinitely long cylindrical warhead and for a sphere, and Sterne's two formulas, developed for a flat layer of explosive with metal plates on one or both sides. Denoting the ratio of explosive charge mass to metal mass in a unit-length cross-section of the warhead as c/m, the initial fragment velocity, V_o, can be found as follows:

$$V_o = \alpha \sqrt{\frac{c/m}{1 + 0.5\, c/m}} \qquad \text{(Gurney, solid cylinder)} \qquad (4\text{-}3.15)$$

$$V_o = \alpha \sqrt{\frac{c/m}{1 + 0.6\, c/m}} \qquad \text{(Gurney, sphere)} \qquad (4\text{-}3.16)$$

$$V_o = \alpha \sqrt{\frac{0.6\, c/m}{1 + 0.2\, c/m + 0.8\, m/c}} \qquad \text{(Sterne, flat plate)} \qquad (4\text{-}3.17)$$

$$V_o = \alpha \sqrt{\frac{c/2m}{1 + c/6m}} \qquad \begin{array}{l}\text{(Sterne, symmetrical flat} \\ \text{sandwich, each} \\ \text{plate of mass } m\text{)}\end{array} \qquad (4\text{-}3.18)$$

where α is a characteristic of the explosive. The derivations of these formulas are based on an assumed distribution of gas velocities, with the gas velocity equal to the fragment velocity at the interface. (See References 4-3.e and 4-3.f.)

To use any of the aforementioned equations, a value of α is required. Theoretically, $\alpha = \sqrt{2E}$ where E is the energy, per unit mass of explosive, convertible to mechanical work. It is to be noted that this is not the same as the total energy of a unit mass of explosive. Indications are that the following values, reported in Reference 4-3.b, are appropriate for solid cylinders.

TNT	α = 8,000 ft/sec
Composition B	α = 8,800 ft/sec
Composition C3	α = 8,800 ft/sec
H-6	α = 8,650 ft/sec

For flat plates, values of α that are 25 percent higher are thought appropriate since the casing of a cylinder ruptures when about

80 percent of the energy E has been converted into mechanical work. It is actually impossible to say what the exact values of ∝ are, since the definition is not susceptible to experimentation and one can only deduce answers from the velocities observed. Thus in a particular geometry there is no clear basis for saying whether the velocity is lower than predicted by the formula or a different value of ∝ should be used to represent the effectiveness of the explosive under those conditions. However, the value of ∝ is probably dependent on the density which the explosive has when loaded.

Generally speaking, the Gurney formula for solid cylinders has given good agreement with experiment for long cylinders (length/diameter or $L/D = 2.5$; in some cases even for $L/D = 1.25$), and moderately good but somewhat high results for short cylinders or ogives, and Sterne's formulas give good results for thin hollow cylinders.

For example, for annular warheads (i.e., with air core) with rather thin layers of explosive and large radii, Sterne's "sandwich" formula (equation 4-3.18) is found by tests to be a good approximation.

To further assist the designer, the fragment initial velocities computed by use of equations 4-3.15, 4-3.16, and 4-3.17 for the solid cylinder, sphere and flat plate are plotted against c/m in Figure 4-15. The curves shown are plotted on the basis of Composition B explosive, and incorporate a table of correction factors for other types. The use of this table will be immediately apparent. The designer is reminded to bear in mind that the initial fragment velocity obtained in this step is a maximum possible value, and is not necessarily the optimum. Note that lower values are found near the ends of a cylindrical warhead and near an edge of a plate warhead.

Optimum Fragment Weight and Velocity--Aerial Targets A fragment's damaging power against aerial targets can be measured by the thickness of metal it can penetrate and by its ability to initiate fires or to damage bombs carried by the target aircraft. Therefore, for any

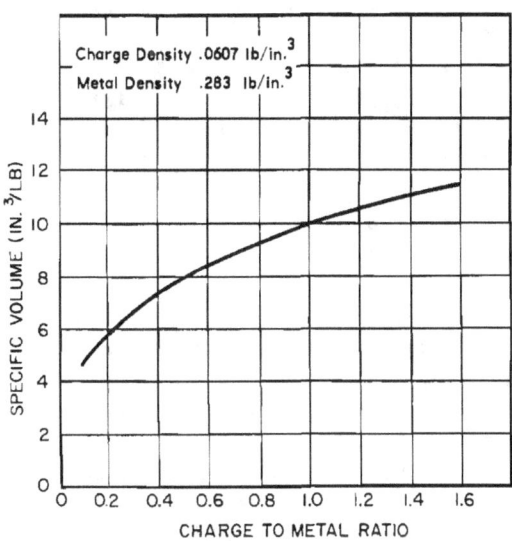

Figure 4-14. Volume Per Pound of Warhead Vs. Charge-to-Metal Ratio

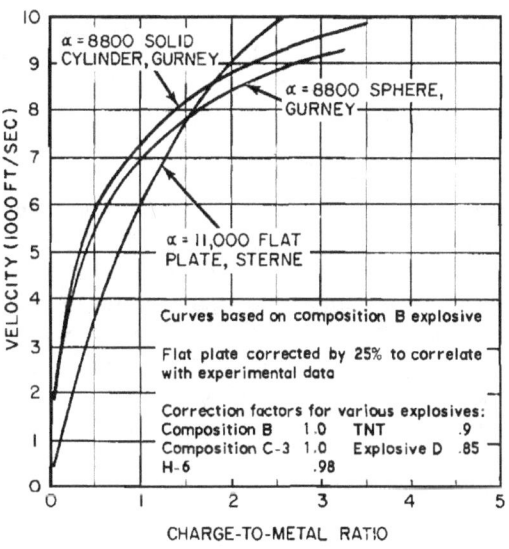

Figure 4-15. Initial Static Fragment Velocity Vs. Charge-to-Metal Ratio

given velocity, in particular for the optimum velocity, there is an optimum fragment size. The optimum is dependent on fragment slowdown, hence on altitude and guidance error.

In choosing the optimum fragment size against aerial targets, there are several other factors which must be considered. Very small fragments which are optimum at high altitude would be nearly useless at low altitude except for very small guidance errors. Since the missile is likely to be required to operate over a range of altitudes, ineffectiveness of a given size of fragment at the minimum required altitude rules out that size as a choice.

The probable target has a major influence on fragment size. For example, in the case of aircraft, the armor thickness around vital components varies considerably between different models. The warhead designer is therefore faced with the problem of selecting the probable thickness of armor of his target. Effectiveness of very small fragments against jet engines has been discounted in most cases.

Another factor which must be considered is that the shape of the curve of warhead effectiveness against fragment size is not symmetrical. It is essentially zero for very small fragments (ignoring blast effects) until the useful threshold is reached, then rises steeply as fragment size increases, has a rather broad maximum, and falls slowly as fragment size increases beyond the optimum. The useful threshold varies according to the target. Figure 4-16 shows qualitatively a typical curve form. For a large departure from optimum fragment size, it is evident that the penalty for choosing too small a fragment size is much greater than the penalty for choosing too large a fragment size. Against engines and bombs the optimum effective fragment size is strongly increased over that for penetrating the skin of the target.

In any analysis of fragment size and velocity based on target skin penetration, rather severe approximations and lengthy calculations are involved. For this reason, graphs of effectiveness of fragments of optimum size and velocity against aerial targets are presented for the convenience of the designer in Figures 4-17 through 4-33. These figures are a very rough guide for use in preliminary design only Reference 4-3.g. These curves of relative effectiveness are based on recent experimental vulnerability data rather than penetration laws and the effectiveness scale is in arbitrary units. The results presented are in general accordance with these penetration laws except for a tendency to require slightly greater fragment size.

References 4-3.x and 4-3.y present detailed experimental data on steel fragment velocity and size needed to penetrate various thicknesses of mild steel (Reference 4-3.x) and armor materials (Reference 4-3.y) at different obliquities. Empirical formulas are presented in conjunction with their graphical representation. Data is included for fragment sizes of from 10 to 1000 grains with velocities of 400 to 6000 feet per second. Reference 4-3.y .

To tentatively select the optimum fragment size and weight, one should refer to the presented curves for a target with characteristics similar to the target of his warhead at the proper altitude and guidance error. The curves may be interpolated for altitudes and guidance errors not presented. Since the maximum initial fragment velocity based on the maximum allowable c/m has previously been established, one can readily find the fragment size and velocity (equal to or less then $V \max$) which will reflect in maximum effectiveness. If the missile must be effective at more than one altitude, as is generally the case, the fragment size-velocity curves should be plotted or transposed on the same curve sheet in order to be certain that the selected fragment size and velocity for the one altitude results in a reasonably near optimum effectiveness for the other altitude in question. If this is not true, the designer should select a combination of fragment size and velocity which will be reasonably near optimum effectiveness for the altitude range desired. An example of this selection process follows.

The designer may have a case where two possible operating altitudes are specified. For example, if the operational characteristics of the missile designate a guidance error of 100 feet against a piston engine fighter at both sea level and 30,000 feet, and the designer has determined from the charge to metal ratio versus velocity curve that his maximum velocity is 6,000 feet per second, the optimization procedure is as follows. Both Figures 4-17 and 4-19 must be used. From Figure 4-19 it is seen that a 0.05 ounce fragment would be best at 30,000 feet. However, from Figure 4-17, a 0.1 ounce fragment is optimum for an altitude of 0 feet. It is obvious from examining the graphs that very little effectiveness would be lost by using a 0.1 ounce fragment at 30,000 feet as compared with attempting to use a 0.05 ounce fragment at 0 feet. Therefore, a 0.1 ounce fragment is the logical choice since it has a fairly high effectiveness at both altitudes. It can also be seen from the curves that in this case the designer should try to maintain the maximum velocity possible as this will lead to the best relative effectiveness. It is possible that, depending on the operating conditions and maximum velocity, reducing the velocity will result in better overall effectiveness, and in cases of operation at more than one altitude all facets of the situation should be considered.

The optimum fragment size and velocity for purposes of causing detonation of aerial targets such as missiles are reported in References 4-3.p and 4-3.q.

Some fragmentation warheads are designed to initiate fires in the target. In the case of aircraft, the primary targets are the fuel cells and fuel lines, and the secondary targets are oil and hydraulic lines, oxygen or acid tanks, etc. It is to be noted that fires can rarely cause "A" or quick kills, but are ideal for causing a "B" kill which results in target destruction in approximately twenty minutes. Fuel lines can be killed by relatively small fragments, but unless all the aircraft fuel lines are interconnected, such damage only causes the loss of a single engine.

When aluminum plate is struck by a steel fragment, the aluminum is pulverized and a flash occurs. The higher the fragment striking velocity, the greater the flash. The flash produced by fragments striking at less than 4000 feet per second is not effective. The thickness of the plate also affects the flash. If the target is thin as on most aerial targets, the flash occurs on the far side of the plate; if the target is thick, the flash occurs on the near side of the plate. Obliquity of the target plate tends to produce larger flashes. The average duration of a fragment produced flash is approximately five milliseconds.

If an aircraft fuel cell is the primary target of the fragment, the type and protection of the cell governs the size and velocity of the fragment. The objective of the fragment is to penetrate the protective plate of the fuel cell and create holes through which the fuel squirts outward. Thus the thickness of the protective armor governs the necessary fragment velocity. Usual velocities for this purpose are 6000 feet per second and higher. The accompanying flash subsequently starts the desired fire. If the fuel cell is of the integral type, fires are very difficult to start and maintain. Generally 2-6 inches clearance between the cell wall and the aircraft structure is necessary to start and promote a fuel fire. In general, a 120 grain fragment is considered the minimum size for creating fires in a self-sealing fuel cell, and a 30 grain fragment is the minimum size for a bladder-type fuel cell.

Data on the effects of altitude on a fragment's ability to initiate fires are somewhat lacking. Fires can be started at altitudes up to approximately 75,000 feet; fires can be started with fragments at altitudes up to approximately 65,000 feet. The flames however, are not as hot or as violent and are less damaging at high altitudes. The lower the ambient temperature, the more difficult it is to start a fire. It is to be noted that the local ambient in aircraft is dependent upon the type of aircraft construction as well as the operating altitude. In general, from 0 to 20,000 feet the fire starting capabilities of a fragment are good, and do not vary. From 20,000 to 35,000 feet the fire starting capabilities are somewhat adversely affected, and from 35,000 feet upward these capabilities are poor and become increasingly worse.

Pyrophoric fragment materials give a better flash than steel fragments, but their use is only justified in the event that penetration of the target is obtained. Titanium and stainless steel targets flash less than aluminum ones, and consequently a higher velocity fragment is required for these targets to produce a comparable flash. In addition, larger fragments are required for penetration of titanium and stainless steel. Honeycomb aluminum structures produce larger flashes than single sheet aluminum structures.

It is to be noted that the foregoing discussion concerning the use of fragments to initiate aircraft fuel fires is relevant to the present fuels of the JP type. The advent of new type fuels could alter the advisability of the use of fragments for such purposes.

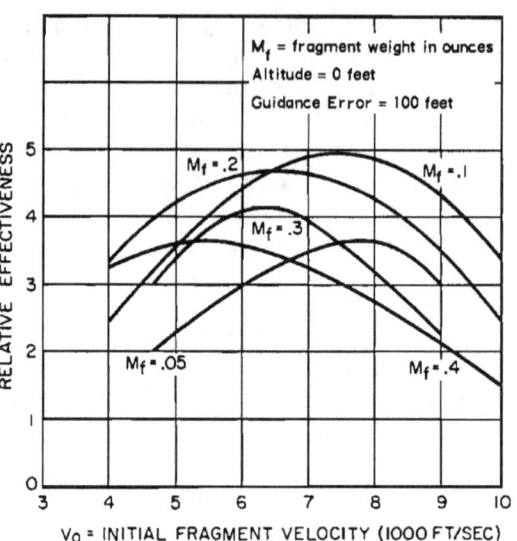

Figure 4-17. *Fragment Velocity and Size Optimization, Target: Piston Engine Fighter*

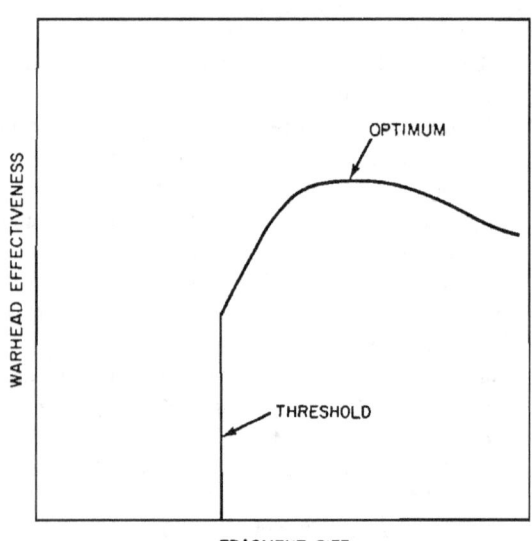

Figure 4-16. *Warhead Effectiveness Vs. Fragment Size*

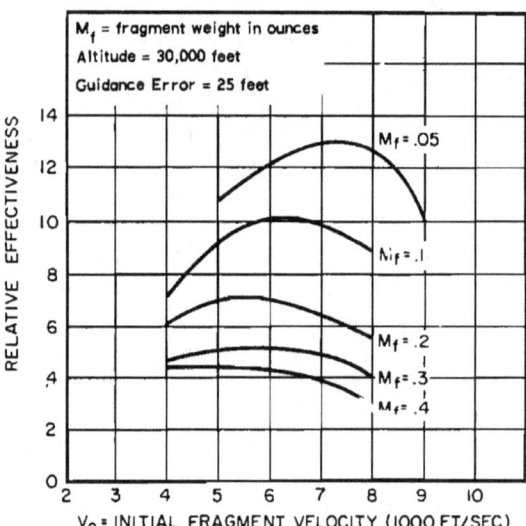

Figure 4-18. *Fragment Velocity and Size Optimization, Target: Piston Engine Fighter*

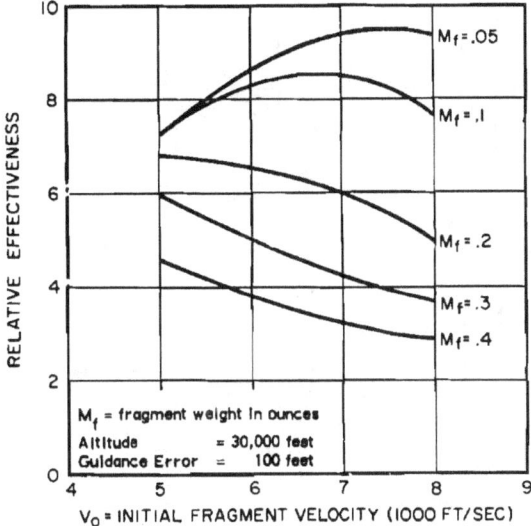

Figure 4-19. Fragment Velocity and Size Optimization, Target: Piston Engine Fighter

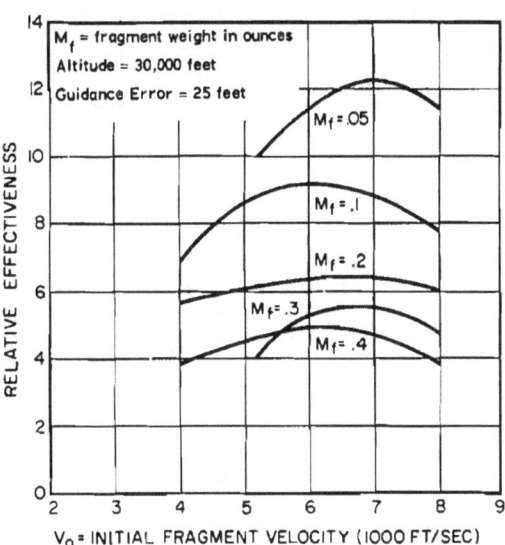

Figure 4-21. Fragment Velocity and Size Optimization, Target: B-29 Aircraft with Fuel

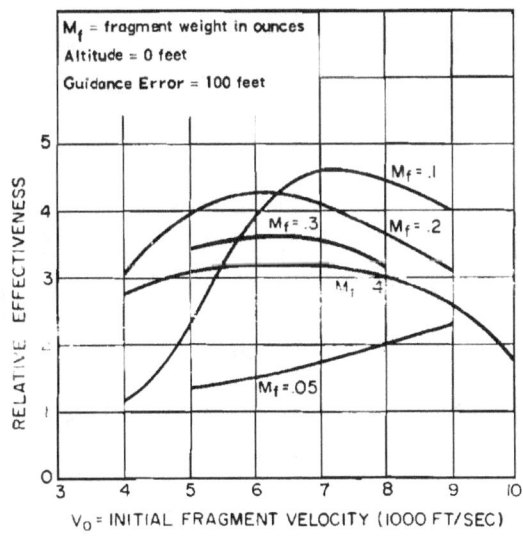

Figure 4-20. Fragment Velocity and Size Optimization, Target: B-29 Aircraft with Fuel

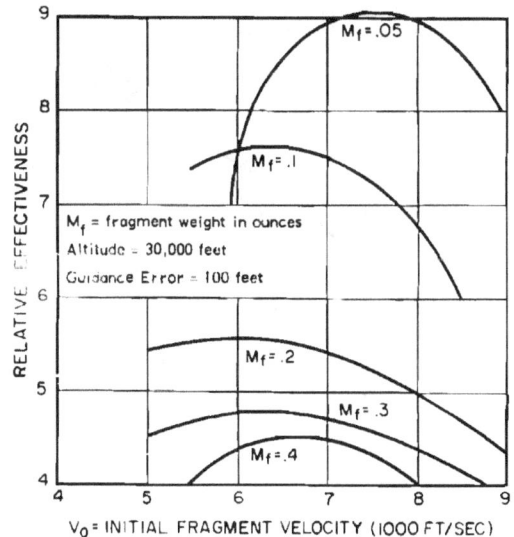

Figure 4-22. Fragment Velocity and Size Optimization, Target: B-29 Aircraft with Fuel

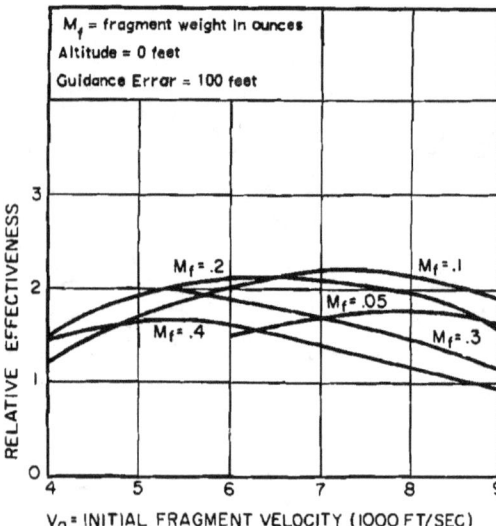

Figure 4-23. Fragment Velocity and Size Optimization, Target: B-29 Aircraft with Fuel Invulnerable

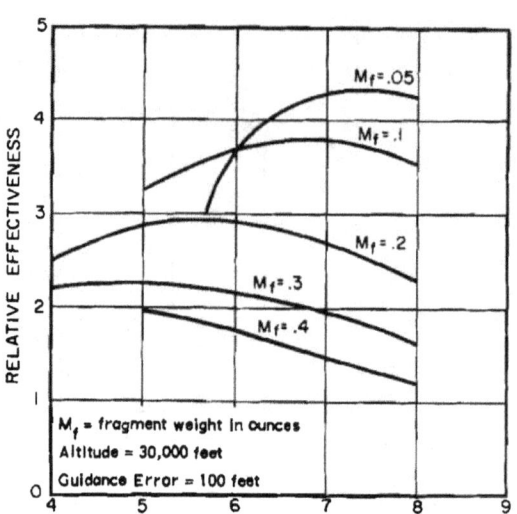

Figure 4-25. Fragment Velocity and Size Optimization, Target: B-29 Aircraft with Fuel Invulnerable

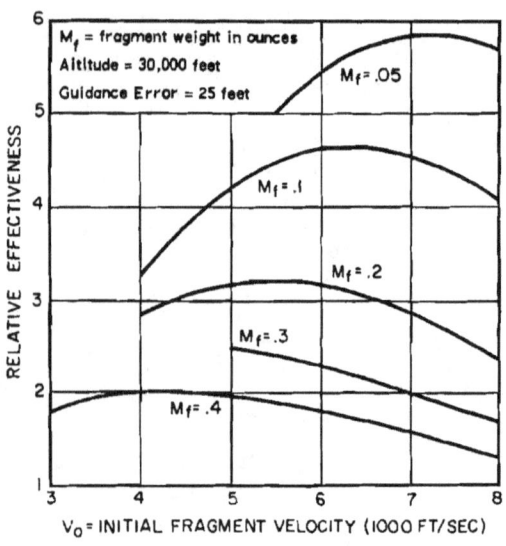

Figure 4-24. Fragment Velocity and Size Optimization, Target: B-29 Aircraft with Fuel Invulnerable

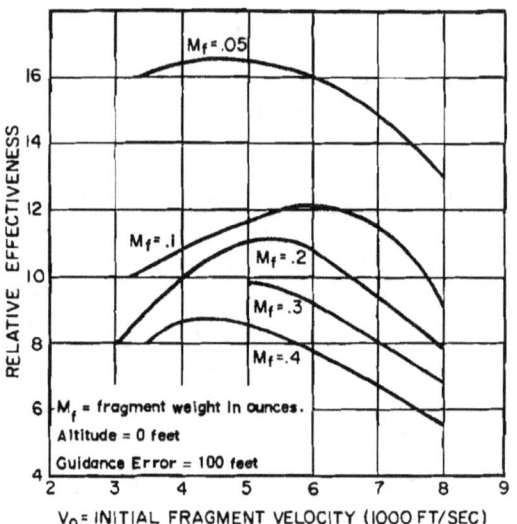

Figure 4-26. Fragment Velocity and Size Optimization, Target: Single Engine Jet Fighter

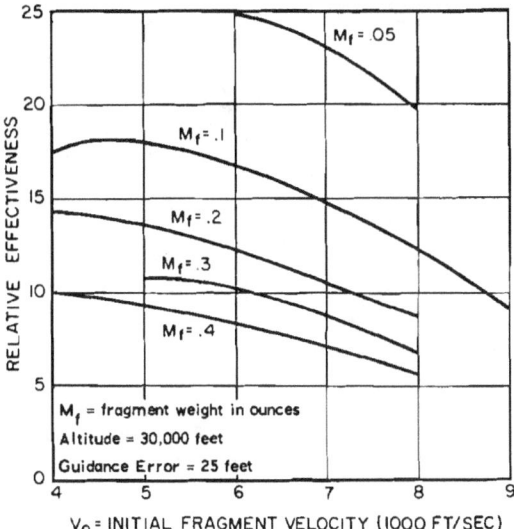

Figure 4-27. Fragment Velocity and Size Optimization, Target: Single Engine Jet Fighter

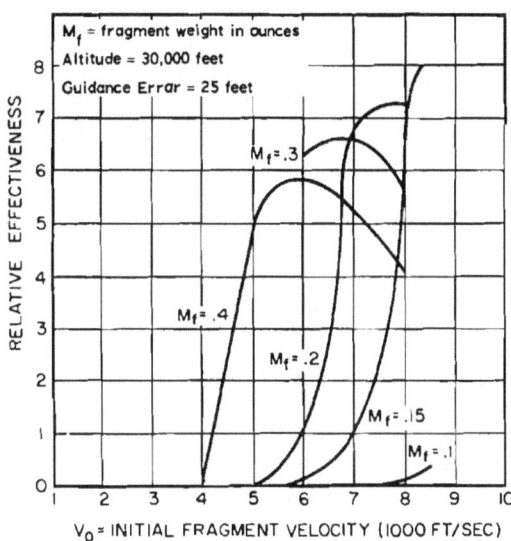

Figure 4-29. Fragment Velocity and Size Optimization, Target: High Explosive Airborne Bomb

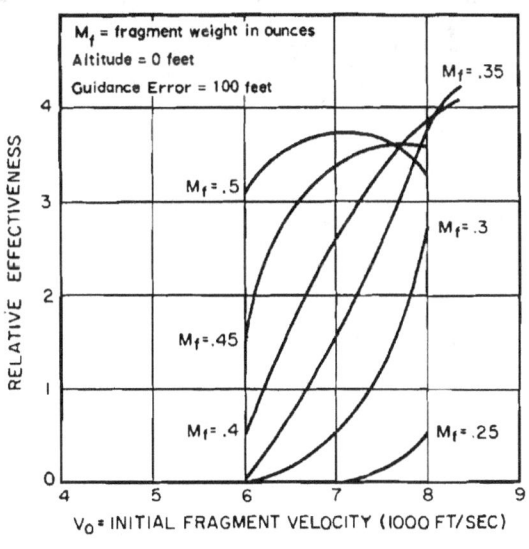

Figure 4-28. Fragment Velocity and Size Optimization, Target: High Explosive Airborne Bomb

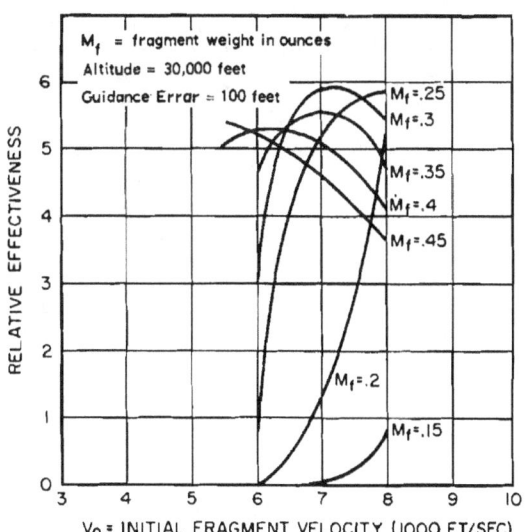

Figure 4-30. Fragment Velocity and Size Optimization, Target: High Explosive Airborne Bomb

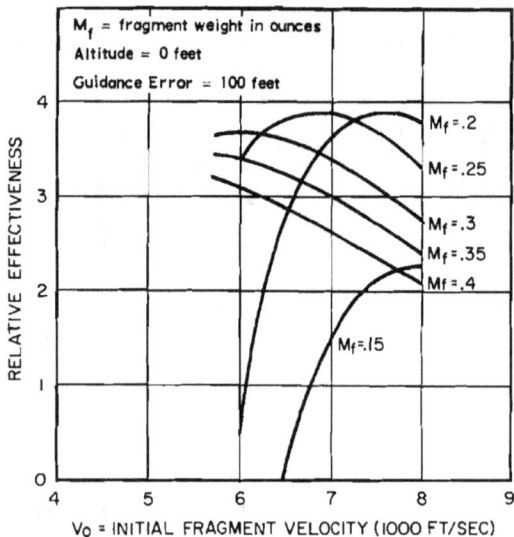

Figure 4-31. Fragment Velocity and Size Optimization, Target: High Explosive Airborne Torpedo

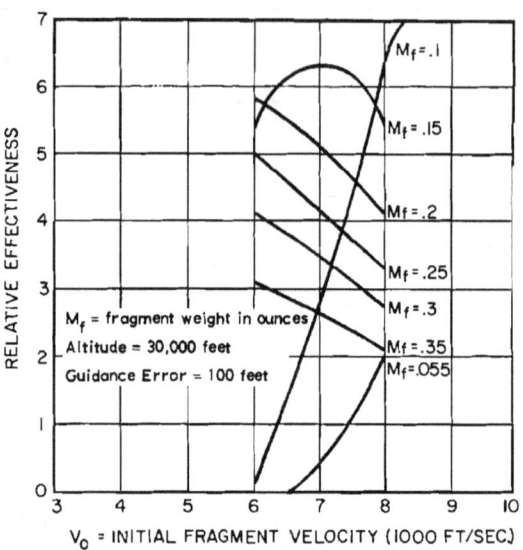

Figure 4-33. Fragment Velocity and Size Optimization, Target: High Explosive Airborne Torpedo

Figure 4-32. Fragment Velocity and Size Optimization, Target: High Explosive Airborne Torpedo

Figure 4-34. Velocity Ratio Vs. Range, Anti-Personnel Warhead

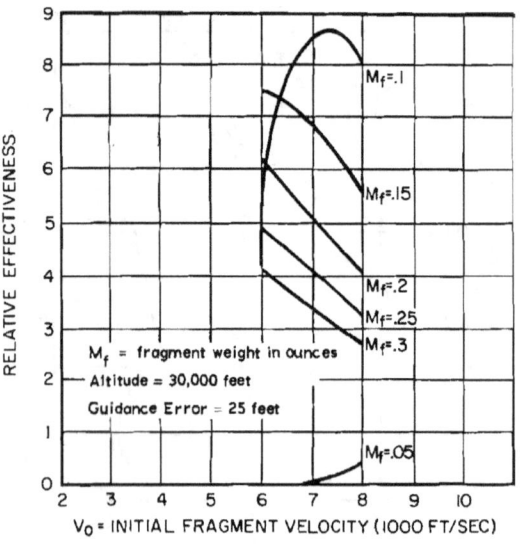

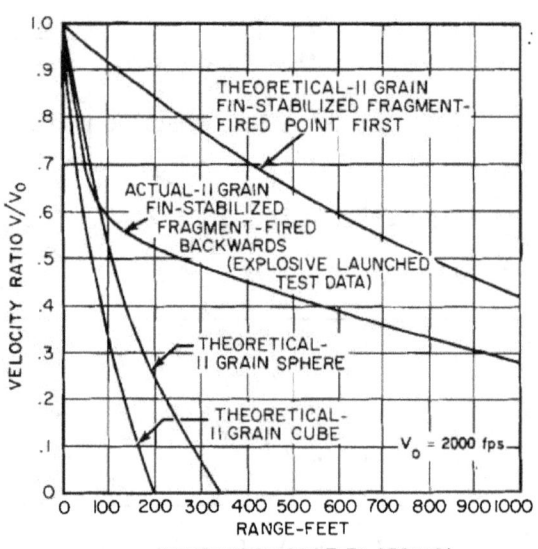

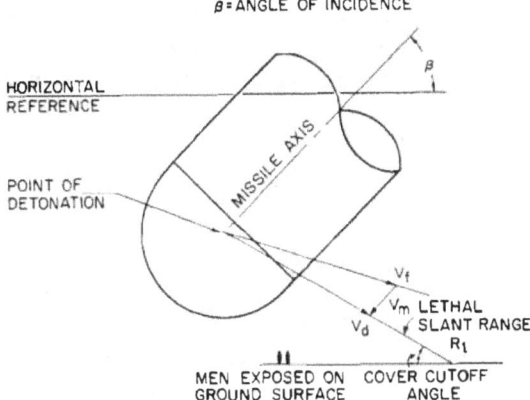

Figure 4-35. Vector Addition of Fragment and Missile Velocities, Anti-Personnel Warhead

Optimum Fragment Weight and Velocity--Ground Targets In the case of anti-personnel warheads, the optimum fragment size and velocity is a function of the type of fragment used. There are two general types: the spherical or cubical fragment, and the fin-stabilized or "needle" type. The fin-stabilized type has better aerodynamic characteristics and inherently better penetration, but has a high initial cost for the experimental item. This type has been developed for a few of the more recent warheads. If development time is sufficient for packaging and if it proves that this type can be launched without excessive breakage, it is recommended that fin-stabilized fragments be considered. However the development time is less for warheads using conventional preformed spheres or nearly cubical fragments, and they should be used in interim warheads.

The optimization of size and striking velocity of the fragments is a field in itself, and opinions vary considerably on the subject. Due to the shape and striking attitude of the fin-stabilized fragment, it is readily apparent that a lesser force or impact energy is necessary for incapacitation than for a spherical or cubical fragment. Also, the smaller the individual fragment, the greater number of fragments are possible in a given volume. (However more spheres or cubes of a given mass can be packaged in a smaller volume than darts or flechettes of the same mass.) Conversely, the smaller the fragment, the higher the velocity required for its effectiveness, and hence the greater c/m ratio and less total weight available for fragments. Although no concrete simplified basis can be given for selecting the optimum fragment size and striking velocity because of the complexity of the damage criteria, warheads have been designed using the following criteria. For the case of the spherical fragment, fragments of 28 to 240 grains (437.5 grains = 1 oz) have been utilized with a striking velocity of from 1000 to 6000 feet per second.

The increased number of potentially lethal or disabling small fragments increases the probability of at least one hit on an individual within the effect-area of the warhead. However the effect-area for small fragments shrinks with decrease of fragment size unless the initial velocity is increased by increasing c/m which reduces the number of lethal fragments. In the end, the optimum warhead is a compromise that also involves burst height and the coverage of personnel in typical positions.

For the latest available data on incapacitation by fragments, one should consult the Contact Wound Ballistics Laboratory, Army Chemical Center, Edgewood, Maryland. If it is not feasible to do this, sufficient data may be found in the work of Allen and Sperrazza, reference 4-3.h, to make possible an estimate of a usable combination of fragment size and velocity.

Knowing the striking velocity desired and distance above the ground at detonation, the c/m required may be calculated as follows. Use can be made of equation 4-3.2 in computing the initial dynamic velocity, as explained previously. The drag coefficient, C_D, is at a maximum in the vicinity of Mach 1, which degrades performance for operation near this point. In addition, the effect of velocity slow-down at sea level is far more pronounced than that at a high altitude because of the relatively high density of the atmosphere at sea level. Figure 4-34 illustrates the effects of the drag on the typical anti-personnel fragment shapes for an 11 grain fragment size. Once the initial dynamic velocity is known, it remains to find the initial static fragment velocity. The initial dynamic velocity (V_d) is a function of both the initial static velocity (V_f) and the velocity of the missile (V_m). If, e.g., the shape of the nose-spray warhead and the minimum angle (θ_s) of incidence to the ground have been previously established, Figure 4-35 may be drawn. As an approximation, the detonation point can be so located that the path of a fragment at the bottom of the rear edge of the warhead will be inclined at angle θ_s, in the notation of Figure 4-10, by drawing a line between the point of detonation and the last-mentioned fragment as shown in Figure 4-35.

Knowing the direction of V_f (θ_s) and the magnitude of V_m and V_d, V_f may be found from equation 4-3.19 which is

$$V_d^2 = V_f^2 + V_m^2 + 2 V_f V_m \cos \theta_s \qquad (4\text{-}3.19)$$

Now, with V_f known, the c/m required to provide it may be found from Figure 4-15, taking into account the lowering of the actual velocity of an edge fragment as compared with the theoretical.

Spherical Model for Missile-Carried Fragmentation Warhead A simple spherical model adequately represents the terminal ballistic geometry for a stationary external blast warhead, as is self-evident. The spherical model can likewise be used for missile (moving) warheads of both the fragmentation and external blast types.

At any early time t after a stationary fragmentation warhead has burst, most of the fragments are at nearly the same distance $R = \overline{V}_f t$ from the warhead position, where $\overline{V}_f$ is the average fragment velocity (considering slow-down due to air-drag) during the time t. Otherwise stated, most of the fragments are near a rapidly expanding spherical surface of radius $R = \overline{V}_f t$ and centered on the warhead position.

Likewise at time t after a burst of a missile fragmentation warhead moving at velocity V_m, a spherical surface of radius $R = \overline{V}_f t$ adjacent to most of the fragments is centered at a point on the missile path which is located at distance $\overline{V}_m t = \overline{V}_d \left(\dfrac{V_m}{V_d}\right) t$ from the burst point, where V_d is obtained from equation 4-3.19 and the air-drag slow-down relation, and $\overline{V}_d t$ is the fragment travel from burst to target. In other words, the center of the rapidly expanding sphere is near the point where the warhead would have been at time t if it had not exploded. This spherical model closely approximates the oblate spheroid that actually exists for moving warheads of either the fragmentation or external blast types. (See Section 4-2.2.)

To be useful, fragments generally move at a velocity V_f which is much higher than that of the warhead-carrying missile, V_m. Hence, for such a warhead, it is adequate in both design and evaluation to take the radius of the spherical fragment-containing skin as $R = \overline{V}_f t$. (An exception is where a missile, e.g. for anti-missile use, throws out a cloud of relatively slow moving pellets or submissiles; their radically different geometry is introductorily treated in Reference 4-3. cc.)

In the following section we consider the geometry for bursts of nose-spray fragmentation, antipersonnel warheads. True side-spray warheads seldom can use as many as 50% of their fragments against a surface target. Likewise, spherical warheads use only about 47% of their fragments. On this basis, these low "efficiencies" are to be remembered when examining the warhead efficiencies in the next section. A side-spray warhead can be taken as the difference between two nose-spray warheads of different spray-angles — another reason for first considering the nose-spray warhead.

Design of a Nose - Spray Warhead for Inclination of the Missile Problems of warhead geometry and design are briefly illustrated in this Section. The purpose is to provide an insight rather than to urge the use of any particular method or values of parameters. In particular, certain values—58 ft lb lethal fragment kinetic energy and 10° foxhole cover cutoff, are used mainly to simplify the treatment; these values are here used in preliminary design without implying that they would be used in final optimization or effectiveness evaluation in the future.

For missiles at steep final inclinations (e.g., $\omega' > 45°$), fragmentation missile warheads of the nose-spray type can be more efficient than those of the side-spray type. (For ogival shell, the lethal area increases with the inclination mainly because of the cover-functions. Missile warheads generally have a much higher ratio of V_f/V_m than shell warheads do. Hence, in use, the shell "side-spray" is vectored forward to approach the nose-spray of missile warheads. In other words, the present remarks on the effects of warhead inclination on the lethality of side-spray missile warheads are not to be applied to shell.) At required long ranges, most missiles have inclinations steep enough to reduce the effectiveness of side-spray warheads to very low values and to increase the effectiveness of nose-spray warheads over that for minimum range. Near the required minimum range, the inclination is still so large for guided missiles (usually around half that at the maximum required range) that the effectiveness of a wide-angle nose-spray warhead generally falls off less drastically than does that of side-spray warheads at maximum range. Direct-fire rockets at minimum ranges are more nearly horizontal, but much of this Section applies to rockets used over a wide band of ranges. For example, take limiting ω's of 30 and 60° and design the warhead spray-angle for $\omega = 60°$. (However, one optimizes the fragment mass M and the charge-to-metal ratio c/m for the mean ω of 45°.)

If one takes an inclination of 10° as the virtual or effective cutoff of cover, one has a dynamic beam angle of $\angle b = 30 + 80 = 110°$ from the 60 and 10° angles. See Figure 4-36. From the missile and fragment velocities, $V_m \approx 1000$ and $V_f \approx 5000$ feet/sec respectively, we find that the static beam angle of the bound of the warhead is $\angle a \approx 120°$. A larger value of $\angle a$ would waste lethal fragments in region "A" by projecting them in paths less inclined than the 10° cover cutoff. A much smaller value of $\angle a$ tends to needlessly increase the average distance that the fragments have to travel (by increasing the burst height) to spray all of the area in which men are exposed within the 10° cutoff circle for the lower burst height with $\angle a$ suited to the 10° cutoff and the 60° inclination. In Figures 4-36 and 4-37, the shading represents areas on the warhead for all fragments, and for the static no-cover cap vectored from the dynamic situation.

For similar (usually steel) fragments of varying size and mass M, the minimum lethal striking velocity V_L is found from some function of the mass. For example: V_L is such that the lethal energy is at least 58 ft lb (which was formerly in wide use for fragments heavier than about 25 grains). For this lethal velocity V_L, the lethal distance R_L can be found for different values of the fragment mass M and of the initial fragment velocity V_f (relative to the warhead) which is a function of c/m. To find R_L, we use the air-drag relation $V_L = V_o e^{-\alpha R_L}$ where

α depends on the fragment shape and mass and V_o is the initial velocity resulting from V_f and V_m. The value of R_L is found at the conical bound of the fragment spray (e.g., at "A") since this is where the value of V_o is smallest for a given c/m. In other words, the missile velocity V_m contributes more to V_o (for fragments projected forward) as their angle from the missile direction decreases, and also the fragment velocity V_f drops near the bound of the spray.

The optimum burst height is at or slightly

Table 4-5

Nose Spray Warhead Characteristics

(1)	(2)*	(3)**	(4)	(5)	(6)	(7)	(8)	(9)
			Spherical Warhead		Nose Warheads Designed for ω			Nose Warhead (60° ω Design)
	No-Cover Cap angle	"Used Area" or $\frac{}{2\pi}$	(3)/2			$\frac{\text{Area Frags.}}{2\pi}$	(3)/(7)	Burst at ω's Below
ω, °	$\angle E$, °	$\frac{\text{Area No-Cover Cap}}{2\pi}$	η_s	$\angle C$, °	$\angle a$, °	$1 + \cos C$	η_n	$\eta_n(60°)$
90	91.4	1.024	.512	88.6	91.4	1.024	1.000	-
75	90.9	1.016	.508	73.6	106.4	1.282	.792	-
60	90.3	.995	.497	60.0	120.0	1.500	.663	.663
45	88.0	.965	.483	45.6	134.4	1.700	.568	.598
30	85.9	.928	.464	32.6	147.4	1.843	.504	.527
15	83.7	.890	.445	20.15	159.85	1.939	.459	-
0	81.4	.850	.425	8.01	171.99	1.990	.427	-

*Approximate values from graphed vectors.

**Values from Reference 9-3.n.

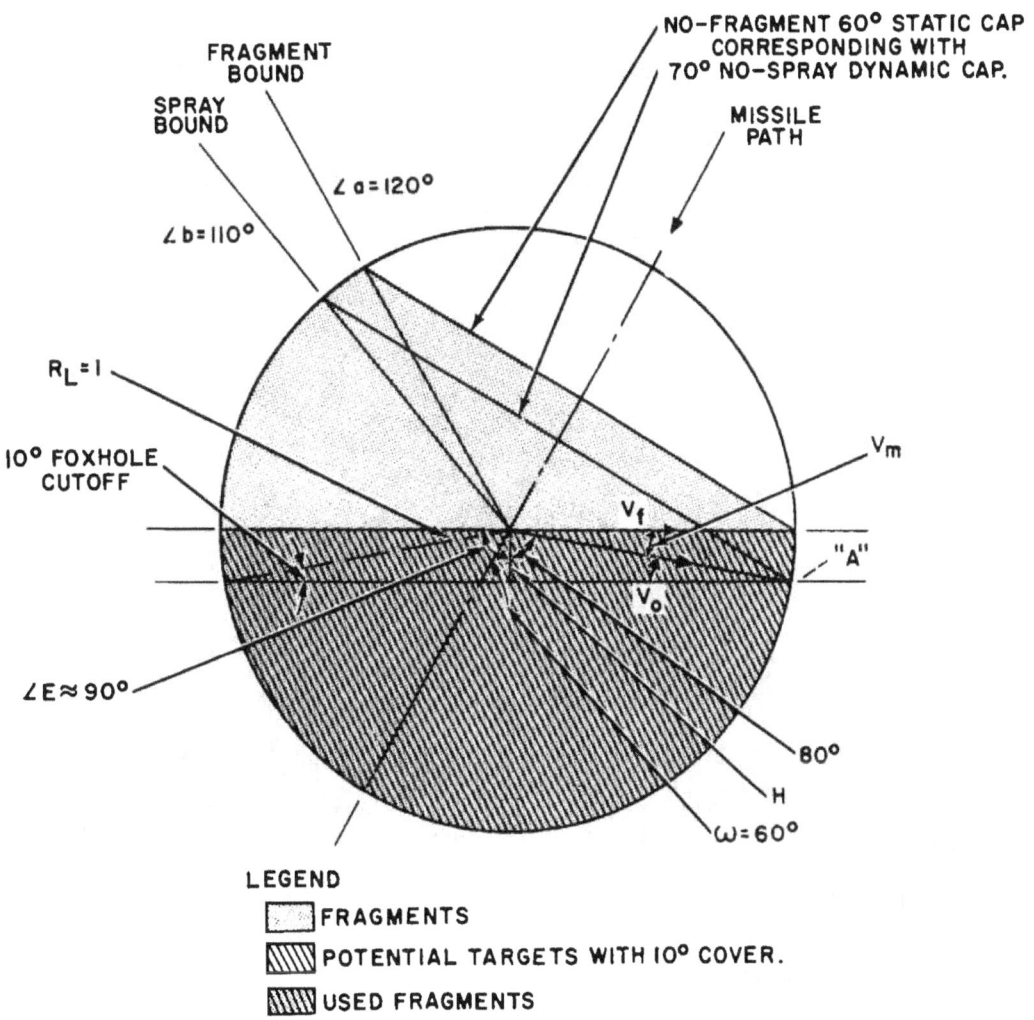

Figure 4-36. *Fragment and Spray Diagram: Unit-Radius Sphere for Burst at $60°\ \omega$ of Warhead Designed for the Same ω*

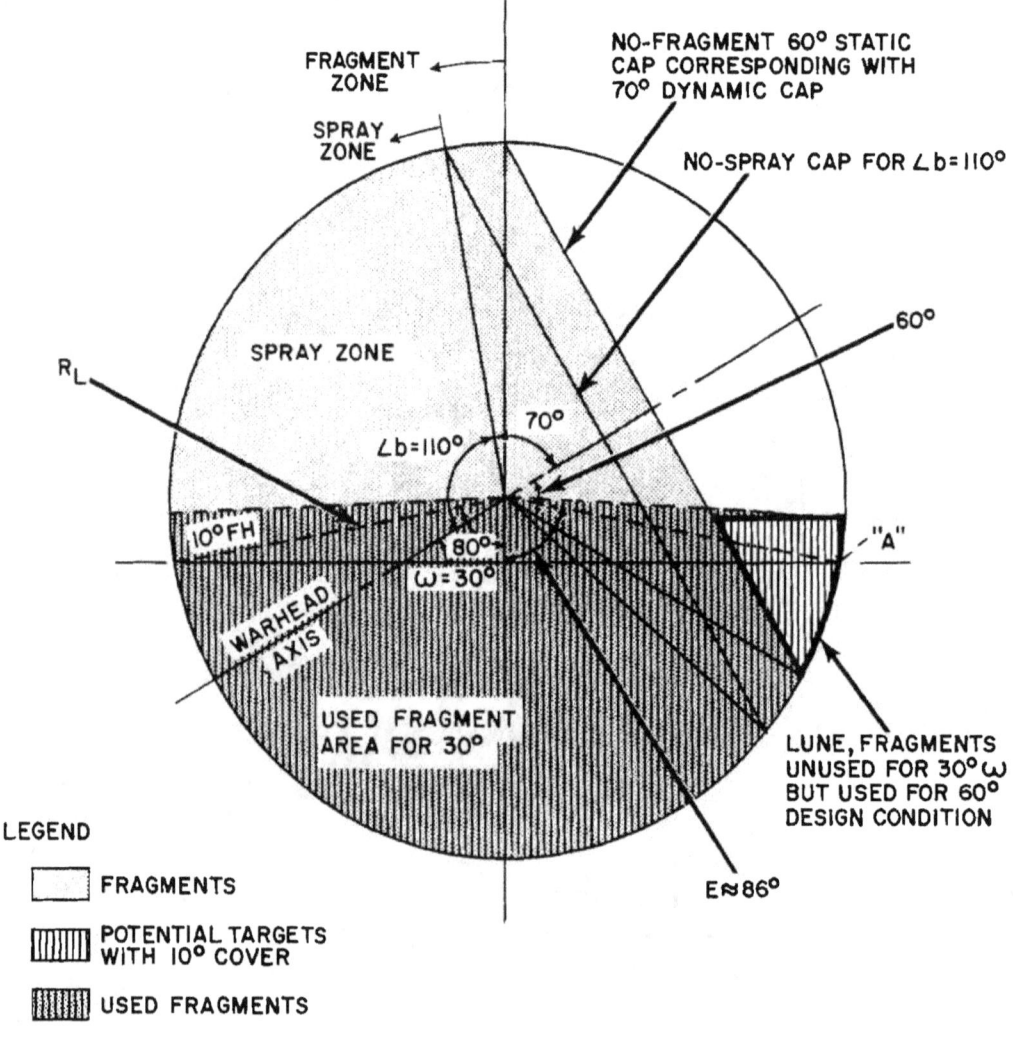

Figure 4-37. *Fragment and Spray Diagram: Unit-Radius Sphere for Burst at $30°\,\omega$ of Warhead Designed for $60°\,\omega$*

above the value $H = R_L \sin 10°$ for this foxhole cover cutoff. The value of H is thus found at ω's of 30, 45 and 60° and the highest value, H_x, of the three is used in the fuzing. For a given ω, the warhead effectiveness falls off rapidly as the burst height drops below H, but negligibly (or not at all) as the burst height exceeds H by a like distance. Hence the fuzing height is set to cause bursts distributed around a mean height H_F that is about $2\sigma_z$ above H_x, where σ_z is the standard fuzing deviation. The use of $2\sigma_z$ causes around 95% of the bursts to occur above H_x.

This simplified procedure, or order of computation steps, is introductory and intended to assist one who only occasionally designs anti-personnel warheads. However, a specialist in this field will probably use more sophisticated methods, especially if a high speed computer is available. In usual practice, one proceeds by a sequential optimization of one parameter after another, with some iteration as maximum performance is approached. But the reader should know that even the best conventional fragmentation warhead cannot closely approach the performance of new types which are already well along in development.

Spray-angles for varying inclination ω and given cover (10° foxhole) will now be discussed. For the 10° foxhole and a nose-type warhead at $\omega > 45°$, the matching static no-fragment angle $C = 180° - \angle a$ is approximately equal to ω. This is apparent from the following table, in which accurate values of $\angle a$ and $\angle C$ are used. In this table, the maximum warhead efficiency η possible was found from the relation $\eta = \dfrac{\text{used area}}{\text{fragmenting area}}$. For a nose-spray warhead:

$$\eta = \frac{2\pi \text{ vers } E}{2\pi(2 - \text{vers } C)}$$

where:

$\angle E$ is the average static angle on the fragment unit-radius sphere, which corresponds with the 80° no-cover cap in the dynamic situation.

In the latter expression, the denominator is the fragmenting area as found from the area of a unit-radius sphere minus the area of the no-fragment zone or cap for $\angle C$. Corresponding values of warhead efficiency η, inclination ω and beam angles $\angle C$ and $\angle a$ are tabulated on Table 4-3.2: (also see Figure 4-37)

For $\omega < 30°$, the fragment area of the warhead increases until the warhead becomes practically spherical. Even for the 30° ω warhead, the fragment zone covers nearly all of the warhead, i.e.:

$$\frac{4\pi - 2\pi \text{ vers } 32.6°}{4\pi} = 1 - 0.5 \times 0.1575 \text{ or } 92.1\%$$

of the spherical area.

For the 10° foxhole and a spherical warhead, the upper bound for the efficiency η can be taken as the fraction of the fragments that reach the ground by linear paths inclined more than 10°. The efficiency relation used is, since $\angle E = 81.4°$,

$$\eta = \frac{2\pi \text{ vers } 81.4°}{4\pi} = \frac{0.850}{2} \text{ or } 42.5\%$$

For the spherical warhead and $V_f/V_m \geqq 5$, efficiency η is nearly independent of the warhead inclination ω, as is shown by the dash curve in Figure 4-38.

For a missile inclination ω of 60° and a warhead that is designed for this inclination and 10° foxhole cutoff, the "used Area" on the unit-radius sphere is that of the 80° no-cover cap. (See Figure 4-36.) The other pertinent zone is the 60° non-fragmenting static cap of this warhead for which the corresponding dynamic 70° no-spray cap just touches the 80° no-cover cap for 60° ω at "A" in Figure 4-36. This 70° (dynamic) no-spray cap corresponds with a 60° (static) no-fragment cap.

However, if the same warhead is in a missile inclined 30°, the 70° no-spray cap takes a lune-shaped "bite" out of the 80° cap of the "used area" for 60°. In other words: for a burst at $\omega = 30°$, the used area (80° no-cover cap) for 60° ω has shrunk because part of it has been replaced by part of the no-spray 70°

cap. (See Figure 4-37.) On the fragment-area basis, the lost lune is bounded on one side by the horizontal 86° cap "no cover" edge-circle and on the other side by the 60° "no-fragment" cap circle which is inclined 60° for the missile inclination of 30°. Since neither of these circles is a great circle on the unit-sphere, the lost lunate area is only about 20% for 30° ω and 10% for 45° ω.

Figure 4-38 presents values of η and ω for both spherical and nose-spray warheads for $V_j/V_m = 5$ and 10° foxhole cover. The dash curve is for spherical warheads. The solid curve is for nose-spray warheads used at the inclinations they were designed for. The three points identified by triangles on this figure are for the nose-spray warhead that was designed for 60° inclination but used at ω's of 30, 45 & 60°. Evidently their slight departures from the design curve are of the order expected for errors in the approximate numerical integration used in obtaining the values of η for these three cases.

Anti-Personnel Warhead Effectiveness The effectiveness of an anti-personnel warhead is ordinarily expressed by its lethal area

$$A_L = \int_0^A p(a)\, da$$ where a is unit-area. (One

can see Reference 4-3.n, Appendix A, for the essential basic relations used in one method of computing lethal areas of fragmentation warheads.) The lethal area A_L is the product of the (ground-area per man) by the (total number of men killed throughout the whole area A that is exposed to potentially lethal fragments). In other words, the lethal area is used to free the expression for warhead effectiveness from the density of men on the ground.

The exposed area of a man in a given position and cover varies significantly with the inclination of the striking angle, i.e., elevation angle of the burst relative to a man. (See Reference 4-3.ee for exposed areas of men standing in the open, Reference 4-3.ff for men in artillery battery positions and in trenches, and Reference 4-3.gg for infantry men in 5 typical defensive positions.)

Near-optimum performance can be produced by a warhead that has a substantially constant fragment density (i.e., fragments per unit solid angle or steradian), as proposed in Reference 4-3.n. Against randomly distributed men and/or fragments, a nose-spray warhead missile is an area-type weapon for which the probability of killing a man within the sprayed area is $p = 1 - e^{-K}$. In this expression, $K = \rho_s A(\theta)$ is the expected number of hits of potentially lethal fragments on a man of projected area $A(\theta)$ in a plane that passes through him and is perpendicular to the fragment path of inclination θ when he is exposed to density ρ_s of the fragments piercing that plane.

The concept of lethal area depends on the assumption that, wherever a warhead bursts, targets are randomly distributed. Many tactical targets are distributed only over areas so limited that the weapon effectiveness cannot be properly expressed by a lethal area (see Reference 4-3.dd, Appendix C); weapon effectiveness is better expressed by either (1) the expected number of targets killed per burst or (2) a sprayed fraction of the target area large enough (usually over 30%) of the target area to neutralize the target area with a given probability that is high (usually over 90%). In general, the first alternative is used for cost estimates and the second for tactical use against important targets that must have their effect eliminated.

In many cases, area-weapons are used against targets of limited area that can be adequately represented by circular or elliptical areas. The Sandia Corporation has extensively treated such targets exposed to weapons having circular areas of effect. Also one can see Reference 4-3.hh for such a target exposed to a small number of bursts of area-weapons. However, the present elementary treatment cannot go further into the more advanced field of weapons systems evaluation.

Actual Charge-to-Metal Ratio and Explosive Type

As previously mentioned, it is quite possible that the optimum velocity for warheads other than anti-personnel will be well below the allowable maximum, and hence a c/m value less than the maximum is indicated. If this is the case, the use of the optimum value will result in a greater metal weight and thereby supply a greater number of fragments. When this is done, the total volume of the warhead will be reduced below the original volume estimated previously.

Selection of the best kind of explosive for the missile warhead still requires extensive study; although of course there is available a great deal of experience with bombs and shells. Composition B has generally been favored for missile warheads used against aerial targets as having satisfactory properties both as to casting and as to detonation. Some new explosives are also bidding for consideration with H-6 apparently in the lead. H-6 has been adopted as the standard nomenclature designation for the composition formerly known as HBX-6. Tests of HBX and H-6 (References 4-3.j and 4-3.k) give slightly lower fragment velocity than Composition B but greater blast effect, while Tritonal gives still lower velocity. These same tests indicated that fragmentation control was about the same for Composition B and HBX, but much poorer for Tritonal.

In the case of antipersonnel warheads, the foregoing explosives discussion is directly applicable if cubical or spherical fragments are used. However, in the case of fin-stabilized fragments, the explosive selected should have a relatively low detonation rate and brisance rating in order to produce more of a pushing than a shattering effect. This latter effect will tend to cause column failure of a dart or damage its fins. An explosive such as Composition D (Ammonium picrate) has been used in some cases. A complete discussion of explosives and their properties is given in the Appendix.

As a first approach, the designer has already assumed the use of Composition B explosive. It is to be noted that the method of design of the fragmentation warhead presented herein is based on optimum fragmentation, and does not attempt to treat the effects of blast which are inherent to some degree in any fragmentation warhead. The ultimate value of this blast effect is most difficult to define, but its effect, especially in cases where sudden kills are required, should not be overlooked. Available data indicate that the blast effect is significant against large targets such as bombers for standard error of guidances up to 30 feet at high altitude and up to 60 feet at sea level. Blast is not highly effective against small targets such as fighters or missiles except at very close range. In the case of anti-personnel warheads, the area sprayed by lethal fragments is relatively so great that the blast effect is of small consequence.

Fragment Shape and Material A cubical fragment is generally preferred against air targets because it has better penetrating power and less drag than an oblong fragment. Moderate departures from cubical shape have only a small adverse effect, so the casing thickness need not be restricted by an exact requirement of cubical shape for the fragments. However, if the charge-to-metal ratio is such that the casing is rather thick, it is generally better to have two or more layers of fragments than one layer. Ultimately the choice of the number of layers is determined by proper fragment shape. It is to be noted that one may judiciously select a fragment size with a minimum sacrifice in effectiveness to obtain a simple shell structure. This would be investigated if the problem of a double walled shell arises.

The foregoing discussion is also applicable to cubical fragments used against ground targets. However, in the case of fin-stabilized fragments the L/D (length/diameter) ratio is most significant. Since the state-of-the-art is such that no firm recommendation can be made as to the optimum value, it is advisable that the L/D of the fragments be checked aerodynamically from a drag and yaw damping viewpoint, and also for possible column failure on ejection. L/D ratios of approximately 10 to 12 have been used successfully. (References 4-3.z and 4-3.aa.)

Little consideration has been given to any fragment material other than steel. Against most components the desire for good fragment penetration argues for a fairly dense fragment material, and it is therefore recommended that steel generally be used. The kind of steel should be selected on the basis of availability and ease of fabrication, since detonation work-hardens soft steel.

Methods of Fragment Size Control It is considered both desirable and practical to control the size of the fragments emitted by the warhead in order to keep to a minimum the amount of metal that will be wasted in fragments too small or too large to be effective. The eventual criterion of successful fragmentation control is not the damaging power of the individual fragment, but of the whole collection of fragments from the warhead; since the warhead weight is usually a prime boundary condition, the number of fragments is, at least roughly, inversely proportional to the weight of the individual fragments. Although a large amount of effort has been expended in studying various methods of controlling fragment size, no one method has been studied sufficiently to provide a really sound basis for final choice between different methods or choice of details of a given method.

Dr. Philip M. Whitman of the Applied Physics Laboratory, Johns Hopkins University, has conducted a thorough study of the various methods of fragment size control (reported in Reference 4-3.1) and this work has been used to form the nucleus of the discussion which follows.

It is to be noted that, although uncontrolled fragmentation is seldom considered for missile warheads, the distribution of fragment size can be predicted by the "Mott Law" (Reference 4-3.r and 4-3.s) which applies to relatively thin casings. This is

$$N(M) = N_f e^{-\sqrt{2M/M_o}} \qquad (4\text{-}3.20)$$

where M_o = mean fragment mass.

Modifications of this formula are available for use if the casing is too thick for the usual Mott Law to hold, i.e., if the breakup is three-dimensional rather than primarily controlled in one direction by the thickness of the casing. (See Reference 4-3.t.) However, this is not usually the condition in missile warheads. Additional work on uncontrolled fragmentation is reported in References 4-3.u through 4-3.w.

Precut Fragments The best method of controlling fragment size is to form or cut the fragments to the desired size before they are installed in the warhead. If this is done, the only possible deviation from the preset size would be caused by breakage upon expulsion, or adhesion to each other or to other parts of the warhead. However, these factors may be considered negligible and for all practical purposes nearly 100 percent fragmentation control is achieved.

This method of control has several major objections which tend to prevent its wide-scale usage. The principal objection is that additional structure is needed for the support of the fragments. This structure usually is formed by a thin metal liner or cover, or both, to which the fragments are fastened with adhesive. This liner, which means additional weight (approximately 10 percent of the total metal weight has been used) contributes little, if anything, to the effectiveness of the warhead. Since weight is of primary concern to the warhead designer, this is a most serious detriment to the use of the precut method of fragment control. It is to be noted that recent developments have proven that plastics such as fiber-glass laminates can be successfully employed as inner and/or outer liners with a definite saving in weight and consequently, more weight can be added in useful fragments or explosive.

An alternate method of installing the preformed fragments, especially spherical or fin-stabilized fragments, is to place them in layers between the inner liner and the case and fill the crevices between them with a matrix to hold them in place and protect them from damage when the missile is fired. A material such as Pittsburgh Plate Glass Selectron No. 5119 (polyester resin) is frequently used for this purpose. In the case of fin-stabilized fragments used in anti-personnel warheads, the fragments may be packed in more than one row, and in various positions, i.e., point first, fin first, etc. They should be packed, however, so as to minimize possible fin damage on explosion.

The primary examples to date of applications of the precut fragment principle were in the fragmentation warheads for the Nike Ajax, Nike Hercules, Hawk, and Bomarc Missiles.

Notched Rings Another method of controlling fragmentation is to form the warhead casing of a series of notched rings fastened together, each forming a section of the warhead perpendicular to the axis of symmetry. This fastening, possibly by brazing, should be considerably weaker than the notched ring material so that breakage will occur where desired. The forces from detonation operate mostly in the direction of stressing each ring circumferentially, and only secondarily to separate adjacent rings. Essentially, the thickness and width of the rings provide control of two dimensions of the fragments, while notches along the circumference of the ring provide places of weakness where breakage in the third direction is desired.

Although this method has been investigated extensively, test results have not been conclusive and the effects of details of the notching has not been finally determined. However, it appears that within reasonable limits good control can be obtained by this method, though possibly only after some trial and error.

It must be kept in mind that the primary purpose of the notches in the rings is to create weak spots in the metal which will fail first after detonation. It would therefore appear that the deeper the notch (within reason) the better the control achieved. However, this is not necessarily true. In some cases very shallow notches have produced excellent results, while in others notches of depth approximately 50 percent of the casing thickness have not given adequate control. Both internal and external notches have been used, and although not conclusive, it appears that internal notches generally give more flexibility of fragment dimensions.

The shape of the notch has received only cursory attention, but in general sharp corners rather than round ones are used since they tend to cause higher stress concentrations which aid in the breakup of the casing. The width of the notch is of secondary importance, and tests indicate that as thin a cut as possible will be sufficient. The spacing of the notches is also indefinite, but it appears that a spacing of 1 to 1.5 times the casing thickness is generally the minimum satisfactory spacing. Radically wider spacing than this (ratio of fragment edges greater than 2 to 1) is generally undesirable because it leads to poor fragment shape for aerodynamic considerations and for target penetration. Also too wide a spacing may result in additional breakup between notches caused by circumferential forces.

In order to minimize the tendency of fragments from adjacent rings to stick together, the notches should be staggered, but the amount of staggering is relatively unimportant; where successive rings have different numbers of fragments because of different diameters, consistent staggering is difficult. Varying the fragment size slightly from ring to ring in this instance is considered more desirable than having some notches aligned. Staggering of the notches tends to produce additional breakage opposite the notches, but not to a serious degree.

Although the material selected for the rings is of relatively minor importance, it must be homogeneous. The material may affect the maximum and minimum sizes of fragments for which control can be achieved, and inhomogeneity can produce erratic results. Test results to date indicate that mild steel might be preferable to high-carbon steel, but the reason for this is not settled.

The method of fastening the rings together is primarily a question of cost and mass producibility, providing the proper strength is obtained. The proper strength may be defined as the strength which will withstand the expected handling and flight loads, but so weak as not to retard the intended breakup. Copper brazing has been the method most commonly considered, but adhesive has given better fragmentation control in some instances with no significant change in velocity and with somewhat less dispersion of the fragment pattern.

Notched Wire Some warhead designs have incorporated notched "wire" wound in a helix, or spiral to control fragmentation. The wire is actually a long bar with two dimensions equal to those desired for the fragments, and is notched at intervals along its length and coiled into the shape of the warhead casing. The wire must be supported by a liner or fastened together by some means (such as welding) in order to preserve the warhead shape. It can readily be seen that if a welding procedure is used, the method of accomplishing the fragment control is basically the same as for the notched ring method. Similarly, if a liner is used, design problems are basically the same as in the precut fragment method. Reasonably good fragment control has been obtained using notched wire; actually, better results have been obtained than in a comparable brazed notched ring warhead.

Grooved Charge The previously discussed methods of controlling fragmentation are substantially similar in that the metal is either precut or notched to cause breakup along predetermined paths. The grooved charge method is the reverse of these. The explosive charge is grooved so that irregularities of the detonation (instead on in the metal) will break up the casing in the desired places. The charge is grooved by means of a fluted liner constructed of plastic, cardboard, balsa wood, or rubber inserted between the solid metal casing and the explosive. When the warhead is detonated, the flutes give a shaped charge effect which tends to cut the metal casing in the pattern formed by the grooves.

The warheads incorporating the grooved charge method of fragmentation control are slightly cheaper and easier to produce than the other types discussed, and there is more flexibility with regard to changing the fragment size. However, this method also has its disadvantages in that there is a loss of some weight and space for explosive and for useful fragments, and there is an addition of some "dead" weight. Test results with a fluted steel liner produced about 14 percent lower fragment velocity than a similar notched wire design, although the flutes gave better fragment control. This loss in velocity is consistent with the difference in c/m caused by the liner. Fragmentation control by this method is probably limited to fragments greater in their lateral dimensions than the thickness of the casing by a factor of approximately 1.2. Further design information can be found in Reference 4-3.m.

Other Methods Various other methods of fragment control have been attempted and subjected to limited testing. Since relatively little data are available on these methods, they will be mentioned only briefly here.

Instead of notching in one direction and having actual discontinuities in the metal in the other direction (such as in the notched ring or wire method), it is possible to cut, punch, or cast a two dimensional network on a solid casing or on pieces later formed and assembled into a casing. Although in principle this method is the same as in the notched rings or wire, preliminary tests gave poor results.

Tests have also been conducted on cast casings with staggered notches, but with no other lines of weakness. This proved to give moderate control of fragment size (about 70 to 75 percent of the weight being ejected in fragments near the design size), and casings heat treated after casting gave somewhat better results than an untreated casing.

Another possibility is to have cases of varying thickness. Tests have been made using casings with humps in the form of segments of a sphere on the inside, with the lines of contact of these segments forming a honeycomb pattern. The results of these tests, although not conclusive, indicate that good fragment control and exceptionally high velocities have been obtained.

The detonation wave could also be shaped by the insertion of inert barriers. This technique is on the borderline between the last mentioned type and the use of grooved charges.

Still another method of fragmentation control is to cast the solid metal casing around wire mesh woven in the desired breakage pattern. The chilling effect of the mesh, and the weakness of the physical discontinuity (especially if the mesh is coated to reduce adhesion) tend to produce breakup in the same pattern as the mesh. This method is desirable because of its simplicity, but to date has not been fully tested.

Comparison of Fragmentation Control Methods

Each method of fragmentation control has advantages and disadvantages whose relative importance to the designer may not be immediately obvious. To further assist in the formation of a design, Tables 4-6 and 4-7 are presented. Reference 4-3.k. In most cases complete information is not available and the factors presented should be regarded as qualitative rather than quantitative.

In some cases the spread between different experimental results or reasonable estimates is so great that the information is presented as a spread, of which the lower end represents results which might well occur with bad luck or inferior design, while the upper end represents what might reasonably be expected in favorable cases.

The method of fragment control has little if any effect on the velocity of the fragments, if it is assumed that any inert material such as liners is counted as (non-valuable) casing weight in the ratio c/m of charge to metal, and account is taken of any explosive displaced. Hence the relative effectiveness of warheads with different types of fragment control is measured largely by the number of fragments of useful size (weighted for dependence of lethality on size) which the warhead produces for a given size and c/m. The relation, however, is not linear, since for some particularly good shots the target will be killed by blast or over-killed by fragments.

Table 4-6 gives a breakdown of the non-explosive parts of the burst warhead; it is assumed that the total of these parts would be substantially the same for all types. Comparison on this basis is slightly unfair to warheads of types requiring structural non-fragmenting members, since they would also require somewhat less explosive to get the same velocity, other things being equal. It is thought that this has been adequately compensated for by taking a conservative estimate of the amount of structure required.

Items such as the metal liners in precut warheads are regarded as chaff and "minor fragments", and are not counted as "structure" in Table 4-6; however, the structure used to carry missile loads through the warhead section is represented as "structural, non-fragmenting".

The significant line in Table 4-6 is that for "relative number of useful fragments". For the various types of controlled fragments, these numbers are in substantially the same ratio as the proportion of weight which goes into useful fragments. For uncontrolled fragmentation, most of the mass goes into fragments of useful size; however, a few fragments are so massive as to drastically reduce the total number of useful fragments (from that of controlled fragments). Although the larger the fragment the more damage it can do, the increase in damage capability is usually far less than proportional to size. It is understood that "useful fragment" is not a clearly defined concept, and that small fragments still have some possibility of inflicting damage in certain cases. The estimates are intended to give partial credit accordingly.

Table 4-7 has as its first row of numbers the estimated relative lethality of the warheads, for equal weights. Although such effectiveness would actually vary somewhat depending on the tactical situation, guidance accuracy, warhead size, fuzing, etc., the given figures are representative of the warhead type. It must be remembered that these numbers are relative to perfect control as unity, and are not intended

Table 4-6

Estimated Relative Fragment Production From
Various Fragmentation Control Methods

	Perfect Control	Uncontrolled	Precut	Notched Rings	Grooved Charge	Cast on Mesh
Relative number of useful fragments	1.0	.3	.8	.5 - .9	.6 - .9	.5 - .8
Proportion of total non-explosive mass to useful fragments	.9	.8	.7	.5 - .8	.6 - .8	.5 - .7
Chaff and minor fragments	0	.1	.1	.4 - .1	.3 - .1	.4 - .2
Non-metal (liners, hot melt, etc.)	0	.01	.01	.01	.02	.01
End plates, fittings, etc.	.1	.1	.1	.1	.1	.1
Structural, non-fragmenting	0	0	.1	0	0	0

Table 4-7

Rough Numerical Comparison of Various
Fragmentation Control Methods

	Perfect Control	Uncontrolled	Precut	Notched Rings & Related Methods	Grooved Charge	Cast on Mesh
Relative lethality	1.0	.7	.9	.75 - .95	.8 - .95	.75 - .09
Relative producibility	0	1.0	.6	.4 - .7*	.7	.6
Relative ease of development from present status	0	1.0	.8	.8 - .5	.7 - .6	.7 - .5

*The higher number might apply to some related methods of manufacture such as welding together notched rings or notching a solid casing.

to be used as precise data. The estimates are based on the relative numbers of useful fragments from Table 4-6, plus allowance for guidance errors so small that either blast damage will occur or the fragment density will be so great that variations in it are unimportant. The expected degree of pattern regularity is taken into account, but no allowance is made for any difference in velocities of the fragments. This was done because, as previously mentioned, the influence of the fragmentation control method on fragment velocity is so very small.

The next line of Table 4-7 "relative producibility", is intended to compare crudely the reciprocal of the cost, which may be interpreted in terms of dollars, or of manhours, machine hours, and materials. This comparison may not be pertinent to a design which will be produced in limited quantities.

The last line of Table 4-7, "relative ease of development", is a rough estimate of the relative amount of effort required, in view of the present status and the inherent difficulties, to develop a satisfactory warhead of a given type for a given weapon. Naturally this also depends in an inverse manner on the degree of perfection sought, as reflected in the lethality.

It is not possible at this time to give a single row of numbers, compounded from all the factors considered in Tables 4-6, and 4-7 which would represent the overall relative merit of the various types of fragmentation control. However, it seems clear that the "relative lethality" is by far the most important of the items discussed, except where one of the other factors is extremely low (such as for the unproducible and unattainable "perfect" control). The various methods should not be compared on the basis of lethality per dollar of warhead cost, since the cost of the warhead is only a small part of the total missile cost.

In view of the inadequate data available to date, no clear-cut conclusions can be drawn from these two figures as presented. It can be concluded, however, that (a) some type of fragmentation control is desirable, and (b) how good a job is done on a given type is probably more important than which type is selected.

Since the conclusions are inadequate to serve as a true guide for selecting a method of fragmentation control, the designer should consider the experience of the manufacturer producing his warhead. Having had previous experience on a certain type may well enable a superior job to be done on that type which will more than overcome the apparently slight theoretical advantages of another type.

Design of Fragmenting Metal Once the c/m ratio, fragment shape, weight and method of control have been established, the detail design of the fragmenting metal can be effected. Using the selected c/m (not necessarily the maximum allowable value originally computed) the total weight of the fragments may be computed as follows:

$$W_m = \frac{W_n}{c/m + 1} \qquad (4-3.21)$$

where:

W_m = Weight of fragmenting metal
W_n = Net weight of warhead

The new net warhead volume may now be computed if the c/m has changed. Using this volume and the warhead shape previously determined to provide the necessary beam width, the surface area of the fragmenting metal should then be computed. Since the individual optimum fragment weight has already been determined, it is now relatively simple to establish the fragment dimensions. It should be kept in mind that the optimum fragment shape (from a packing standpoint) for use against aerial targets is a cube. However, the optimum shape of a fragment may be another shape, e.g., a sphere, which has less air-drag.

In most controlled fragment warheads the inner surface of the fragmenting case is coated with a material known as "cavity hot melt", or "acid proof black paint", which is an asphaltic material similar to that used on roofs. This coating varies in thickness from approximately 1/16 to 1/32 inches and is applied before the explosive is loaded. Its functions are to prevent contact between the explosive and sharp edges of the case to prevent chemical action between the two, to provide some degree of thermal insulation, to effect more uniform case breakup, to fill in crevices in the casing where explosive might be pinched if the case were strained and to provide a bond between the explosive and the metal casing. The type of hot melt used is a function of the type of explosive. Special paints have also been used for this purpose. For example, acid-proof paint, Specification JAN-P-450(2) has been used in conjunction with most standard military explosives.

Design of Warhead Components Other Than Charge and Fragmenting Metal At this stage of the design, one is ready to establish the details of the warhead. To properly accomplish this a detailed structural analysis must be made of the individual components and of the warhead as a unit. The structural design criteria are established by the missile requirements, and are normally given to the designer. It is of the utmost importance that the parts of the missile fore and aft of the warhead be kept in the proper position with respect to each other despite aerodynamic and acceleration loads. Either the warhead itself must be strong enough to perform this function, or additional structural members must be provided. It is most desirable (but not always practical) to avoid distortions of stresses which might crack the explosive, as cracks adversely affect the uniformity and reproducibility of detonations.

The ends of the warhead must be closed to support and protect the explosive. The metal which does this is known as an end plate, and also serves to prevent the explosion gases from simply rushing out open ends instead of accelerating the fragmenting metal. However, the end plates should not be thickened beyond what is necessary to support the charge or provide structural rigidity, for additional explosive will probably do more to confine the main detonation than an equal weight of metal. Accordingly, something between 1/16 inch and 1/8 inch of steel is reasonable to use, although in some cases end plates as thick as 1/4 inch have been used. The optimum thickness can only be determined by test of the particular warhead.

At least one end plate or a central portion of an end plate should be removable to allow for filling the warhead with explosive. This is usually done by pouring for large scale production since it is more convenient, though pressing is used in some cases. In either event the end plate should be either bolted in place, or secured with bayonet-type fittings. The number of bolts required for this purpose should be kept to a minimum.

In cases where the warhead surface does not constitute an external surface of the missile, a fairing (sometimes called windshield) must be provided to maintain the aerodynamic contour. A typical fairing is shown in Figure 4-39. The fairing is usually of aluminum and made as light as possible to minimize the requirement for the warhead to "shoot its way out of its own missile". This fairing is normally supplied by the missile manufacturer, and is not the responsibility of the warhead designer. In the event that aerodynamic heating of the warhead compartment in flight becomes a problem, Rubatex or fiberglass insulation may be applied to the inner surface of the missile skin around the warhead.

The fairing may be attached to the warhead itself or be a structural member. The decision as to whether or not to attach the fairing to the warhead should be predicated on the results of the structural analysis. If such an attachment is necessary, it should be such that it will easily be blown off by the detonation of the warhead.

The required fittings for attaching the warhead to the missile are usually designated by the missile system designer. The fittings must be designed to mate with those in the missile and checked to insure their structural integrity. The position of these attachments will most probably be dictated by the position of the mating missile parts. They are usually attached to either the warhead end plates or the fairing. The detail design of the attachments will de-

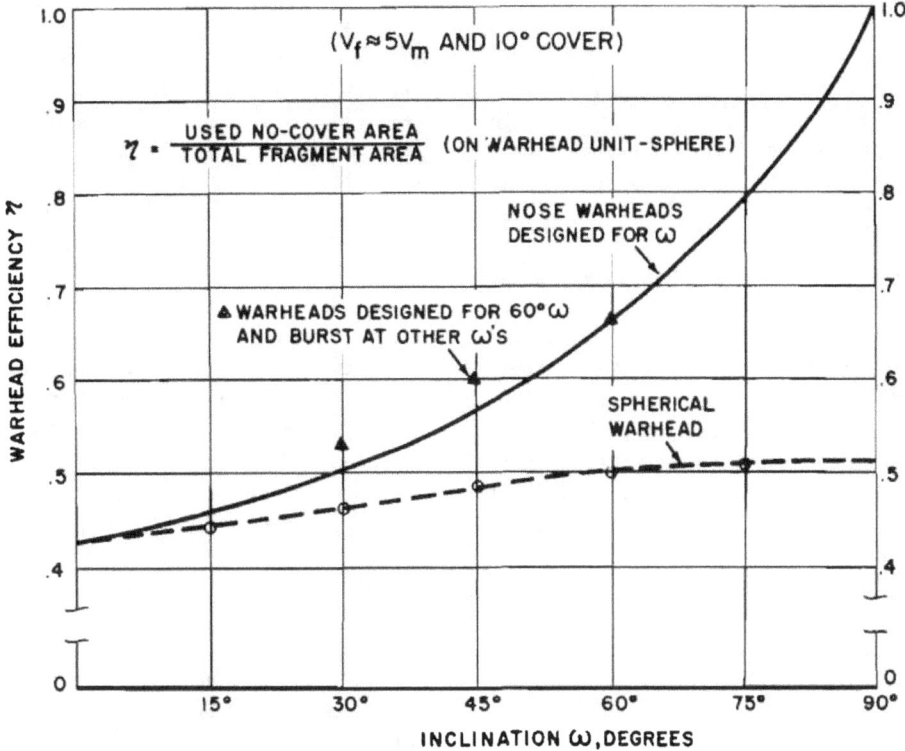

Figure 4-38. Warhead Efficiency Vs. Warhead Inclination

pend on whether or not the stresses are transmitted through the warhead, on the stress level involved, and on whether quick assembly is required.

Handling hooks should be incorporated for use in installing, removing or transporting the warhead. It might be well to check existing handling equipment which will be used to ascertain the compatibility of the design.

It is of the utmost importance to have as little as possible outside the warhead in the way of structural members, wiring, etc. The location of these items is generally specified by the missile system designer. External wiring in the warhead section is a primary source of trouble and should be avoided whenever possible. Tests of warheads with this type of wiring have resulted in numerous failures. It is therefore more desirable, although not always possible, to leave a small conduit down the axis of the warhead for wiring, if wiring is required past the warhead section.

This conduit should be kept small to avoid loss of velocity for the fragments.

The problems of location and mounting provisions for the detonator are of major importance. It is suggested that either fuzing experts or Reference 4-3.1 be consulted to determine the type of detonator to be used. Once the detonator is selected, the necessary mounting provisions and space allotment will naturally be known. The location of the detonator is optional, but a symmetrical location (on the warhead axis) is definitely desirable. In many instances, due to safety regulations, the detonator is mounted within a fuze or S & A device in a safe position so that the explosive train is out of line until mechanical and electrical arming is completed to bring the detonator into line with the explosive train. The location will be determined by the ease of assembly, and by the effect of this location on the fragment pattern. The detonator and associated components are usually located on one end of the warhead or, in the case of very large warheads (over 500 pounds), on both ends.

The location of the center of gravity of the warhead is of great importance to the missile designer, and is specified in most instances. This is especially true of a warhead being designed for an existing missile. After the detail design has been completed, the location of the warhead c.g. should be ascertained. In the event that the c.g. is very critical, the use of ballasting plates is recommended. These plates are usually bolted to the warhead end plates and may either be flush with them or protrude back or forward into the missile. Adjustments can be made by removal of ballasting plates, as necessary. The size and weight of the individual plates are optional. Thus it is possible to compensate for greater variation in the metal parts assembly along with possible changes in the loading density of the explosive, thereby allowing fine control of the location of the warhead c.g.

The detail design of the warhead has now been made. It is to be noted that all assumptions, not heretofore checked, should be verified at this point, and any necessary changes made. The succeeding sections will detail the proper method of presenting the design and the information necessary for assuring the proper coordination of the warhead and its fuze.

Summary of Fuzing Requirements Once the design is final, a summary data sheet should be prepared for the benefit of the fuze designer to permit him to effect a fuze design which will be compatible with the warhead. The following data are required:

(1) Static and dynamic beam width.
(2) Fragment initial ejection velocity.
(3) Fragment size and shape.
(4) A drawing of the warhead.
(5) Type of explosive used.

Summary of Design Data At the conclusion of the design procedure one should prepare a summary of all the pertinent data evolved. This should include the following items:

(1) Total weight
(2) Detail design and installation drawings
(3) Explosive
   (a) Material
   (b) Weight
   (c) Density
(4) Charge to Metal Ratio
(5) Fragments
   (a) Number
   (b) Total weight
   (c) Individual Fragment Weight
   (d) Design Size and Shape
   (e) Initial Velocity
   (f) Beam Width and Beam Axis
   (g) Expected Spacial Density Distribution of Fragments with respect to angle from nose
(6) Location of c.g.
(7) Materials
   (a) Casing
   (b) End Plates
   (c) Fragments

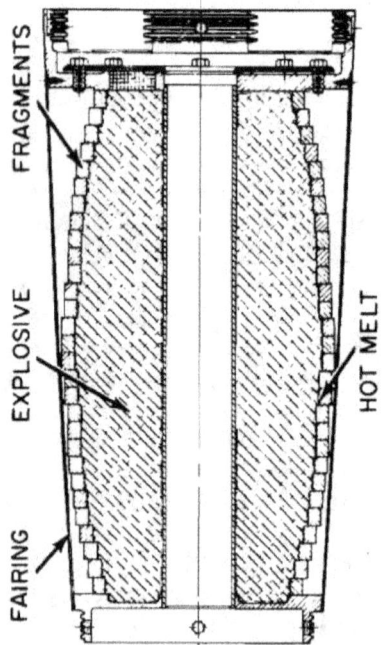

Figure 4-39. Typical Use of a Fairing (Sparrow I, Mk 7 Mod 0, Warhead Shown)

(8) Point of Initiation
(9) Method of Mounting

4-3.3. References

4-3.a "A Measurement of the Drag Coefficient of High Velocity Fragments", J. E. Shaw, BRL Report 744, October 1950.

4-3.b "Guidebook to Anti-Aircraft Guided Missile Warheads, Relation of Characteristics of Warhead and Fragments", Philip M. Whitman, APL/JHU CF2392, June 1955.

4-3.c "A Report on Analysis of the Distribution of Perforating Fragments for the 90MM M71, Fuzed T74E6, Bursting Charge TNT", Harold N. Shapiro, University of New Mexico, UNM/T-234, undated (about 1944).

4-3.d "The Influence of the Surface Contour of an Exploding Body on Fragment Distribution", Willard R. Benson, ASTIA - AD 44460, October 1954.

4-3.e "A Note on the Initial Velocities of Fragments from Warheads", Theodore E. Sterne, BRL Report 648, September 1947.

4-3.f "The Initial Velocities of Fragments from Bombs, Shells and Grenades", Ronald W. Gurney, BRL Report 405, September 1943.

4-3.g "Guidebook to Anti-Aircraft Guided Missile Warheads, Selection of Fragment Warhead Parameters", Philip M. Whitman, APL/JHU CF-2448, November 1955.

4-3.h "New Casualty Criteria for Wounding By Fragments", F. Allen and J. Sperazza, BRL Report 996, October 1956.

4-3.i "Characteristics of Fragmentation of MX 904 Warheads, Blast with Fluted Liner, Composition B, HBX, and Tritonal Loaded", C. L. Grabarek, BRL Memo. Report 700, July 1953.

4-3.j "Comparison of Blast from Terrier Warheads Loaded with Composition B and H-6", Daniel K. Parks, Denver Research Institute, University of Denver, December 1954.

4-3.k "Guidebook to Anti-Aircraft Guided Missile Warheads, Detailed Design of Fragmentation Warheads", Philip M. Whitman, APL/JHU CF-2393, June 1955.

4-3.l "Ordnance Explosive Train Designers' Handbook", April 1952.

4-3.m "British Report on Fluted Liners", ARE/WRD-R55/54, December 1954.

4-3.n "The 1500 lb Anti-Personnel Warhead for the Honest John Rocket", Ed S. Smith, A. K. Eittrein and W. L. Stubbs, BRL Memo. Report 779, April 1954.

4-3.o "The Shape of a Fragmentation Bomb to Produce Uniform Fragment Densities on the Ground", Robert H. Kent, BRL Report 762, June 1951.

4-3.p "Velocity Loss of Projectile Passing Through HE", Jamison and Williams, BRL Memo. Report 1127, October 1957.

4-3.q "High Order Initiation of Two Military Explosives by Projectile Impact", Slade and J. Dewey, BRL Report 1021, July 1957.

4-3.r "A Theory of Fragmentation", N. F. Mott and E. H. Linfoot, Advisory Council on Scientific Research and Technical Development (British), A. C. 3348, January 1943.

4-3.s "A Theory of the Fragmentation of Shells and Bombs", N. F. Mott, Advisory Council of Scientific Research and Technical Development (British), A. C. 4035, May 1943.

4-3.t "Structural Defense, 1945", D. G. Christopherson, Ministry of Home Security, Research and Experiments Department, January 1946.

4-3.u "Experiments with Shrapnel and Wire Wound Model Warheads", S. S. Share, BRL Report 614, July 1946.

4-3.v "The Mass Distribution of Fragments from Bombs, Shell, and Grenades", R. W. Gurney and J. N. Sarmousakis, BRL Report 448, February 1944.

4-3.w "Effect of End Confinement and Shell Length on Spatial Distribution and Velocity of Fragments", Silvia Sewell, BRL Memo. Report 862, January 1955.

4-3.x "Penetration of Mild Steel by Bomb Fragments", N. A. Tolch and A. V. Bushkovitch, BRL Report 568, August 1945.

4-3.y "A Comparison of Various Materials in their Resistance to Perforation by Steel Fragments; Empirical Relationships", Project Thor Technical Report No. 25, July 1956.

4-3.z "Airdrag on Steel Darts and Balls from

Antiaircraft Warheads", Ed S. Smith, BRL Tech. Note 1179, March 1958.

4-3.aa "Retardation and Velocity Histories of an 8-grain Flechette" (or dart), Maynard J. Piddington, BRL Memo. Report 1140, April 1958.

4-3.bb "Optimum Warheads and Burst Points for BOMARC, Phase II Guided Missiles", Ed S. Smith, et al., BRL Memo. Report 739, November 1953.

4-3.cc "Effectiveness of Missile Warheads Against High-Speed Air Targets", Ed S. Smith, BRL Memo. Report 858, December 1954.

4-3.dd "Complex of Soviet Ground Targets on a Stabilized Front", W. A. McKean and Ed S. Smith, BRL Memo. Report 855, December 1954.

4-3.ee "Ground Cover Function for Standing Men Targets", J. W. Marvin, BRL Memo. Report 1089, July 1957.

4-3.ff "Exposure to Airburst Warheads of Men in an Artillery Battery and in Infantry Positions", Ed S. Smith, BRL Memo. Report 1115, November 1957.

4-3.gg "Exposure to Air Burst Weapons of Occupants of Five Specific Types of Defensive Infantry Positions", W. E. Gross, Jr., BRL Memo. Report 1067, March 1957.

4-3.hh "Expected Coverage of a Circular Target with a Salvo of N Area Kill Weapons", Arthur D. Groves, BRL Memo. Report 1084, July 1957.

4-3.4. Bibliography

(1) "Theory of the Explosion of Cased Charges of Simple Shape", L. H. Thomas, BRL Report 475, July 1944.

(2) "Guidebook to Anti-Aircraft Guided Missile Warheads", Philip M. Whitman, Parts I through XII, 1956.

(3) "T-39 and T-40 Missile Warheads", Rheem Manufacturing Company, Monthly and Summary Reports, R 75 Series, June 1953 through March 1956.

(4) "Comparative Fragmentation Tests of Single and Multi-Walled Cylindrical Warheads", Chester L. Grabarek, ASTIA AD 64546, January 1955.

(5) "Explosives Comparison for Fragmentation Effectiveness", A. D. Solem, N. Shapiro, B. N. Singleton, Jr., ASTIA AD 40095, August 1953.

4-4. DISCRETE ROD WARHEADS

4-4.1. Detail Design Steps

Step No.
1. Determine Rod Length
2. Select Rod Material
3. Determine Rod Cross Section
4. Study Rod Velocity Required
5. Select Type of Explosive Charge
6. Design Warhead Details
7. Prepare Summary of Fuzing Requirements
8. Prepare Summary of Design Data

The exact order of the design procedure may vary depending upon the viewpoint of the designer and, even more, on the military requirements which often fix certain parameters in advance.

4-4.2. Detail Design Data

Rod Length One consideration which establishes an upper limit on the length of a discrete rod is quite obviously the length of the compartment allocated to the warhead by the weapon system designer. The lower limit on rod length is established by the damage-producing ability of the rod, that is, it should not be so short that it does not produce critical damage on its intended target, i.e., too short to have a high probability of straddling a structural member. This problem has been studied extensively in Reference 4-4.a. Generally speaking, a 15 inch rod length is the minimum which will cause failure by buckling of the bottom of a B-29 fuselage while 36 inches is the minimum length which will cause a tension failure on top of the same fuselage. Data relating rod length and target cylinder radius to half-sever the cylinder is represented in Figure 4-40 as a guide for establishing minimum rod length for damaging various diameter target fuselages. Data presenting rod length required to half-sever a 3 x 8 foot aluminum beam intended to simulate an aircraft wing is presented in Figure 4-41. These data were taken from Reference 4-4.a. This reference includes a great deal of both analytical and experimental data which is presented in a very orderly manner. The designer is urged to use this reference extensively throughout this and all other steps in designing discrete rod warheads.

The designer will find that the rods required to inflict critical damage to typical aerial targets must be quite long when derived using Figures 4-40 and 4-41, and Reference 4-4.a. Even if his warhead compartment is long enough to accomodate the length established from this damage criteria, he usually will not be able to use this length because long rods have a tendency to break up upon detonation of the explosive charge which propels them outward from the warhead. Generally speaking, rod breakup upon detonation is a serious problem when the length to diameter ratio of the rods exceeds 30. This places a very serious limitation on the length of the rods. Various techniques have been tried to minimize rod breakup such as placing paste or cork liners between the rods and the explosive charge to minimize the explosive shock. It is this limitation on the length of discrete rods which undoubtedly led to the development of continuous rod warheads which are discussed in the next subchapter.

Rod Material Rods inflict damage on aerial targets primarily by severing structure. Light structure such as skin and stringers is severed by direct impingement of the rods or by the overhanging ends of the rods which strike solid structure (such as spar caps), cutting the skin and stringers adjacent to the solid structure. Alloy steels are most effective in this latter case although the rods must not shatter when they strike the structure. Carbon steels, heat-treated to a hardness of about 300 on the Vickers scale provide this capability. For cutting heavy structures such as spar caps, the strength of the rod is less important since their cutting action in this case depends primarily on the rod segment which contacts the target and produces a momentum exchange between the rod and the cap. Mild steel rods are effective against such heavy structures.

Rod Cross Section The cross sectional dimension of the discrete rods generally must be greater than 3/8 inch to be effective against most aerial targets. A higher cross sectional dimension of from 1/2 to 5/8 inch is most desirable. Rods with circular cross sections are not as effective in cutting skin and stringers as are those with square cross sections. The cross sectional shape has little influence on their ability to cut heavy structure such as wing spar caps.

Rod Velocity The velocity with which the rod strikes the target is a function of the vector sum of the missile velocity and the velocity induced by the explosive charge. Striking velocities in excess of 2000 feet per second are required to cause cutting damage to most aerial targets. However, striking velocities considerably in excess of 2000 feet per second will cause other than cutting damage to target components. Vaporific damage caused by the flash induced as the steel rod strikes aluminum in the target occurs at striking velocities of 4000 feet per second and above. The probability of causing vapor damage varies inversely as the volume in which the energy is released. At this time, vaporific damage must be taken as a "bonus" effect. The hydraulic impulse induced when a rod strikes a large body of liquid such as a fuel tank can produce significant target damage. Damage from this source is slight at striking velocities of 2000 feet per second but it increases with velocity until at 6000 feet per second it can be responsible for an immediate kill.

Explosive Charge The type and weight of explosive to be used in discrete rod warheads is very difficult to set forth due to the many variables involved and more particularly, the fact that the prediction of rod breakup upon deto-

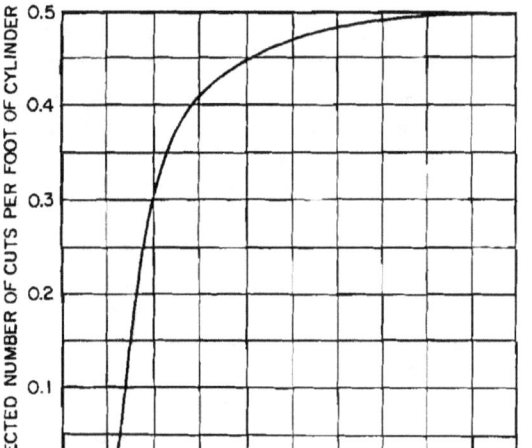

Figure 4-40. Expected Number of Cuts Vs. Rod Length (Cylinder Half Severed)

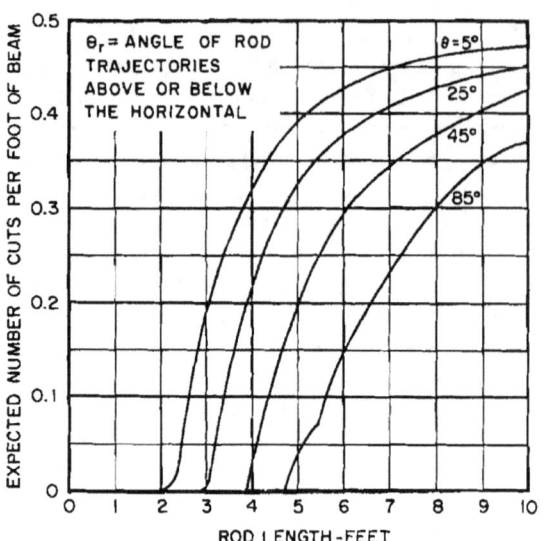

Figure 4-41. Expected Number of Cuts Vs. Rod Length (3' x 8' Beam Half Severed)

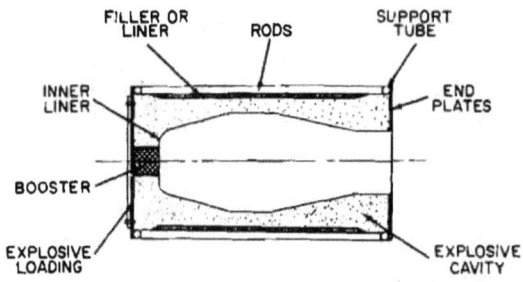

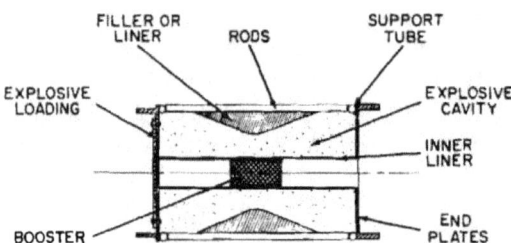

Figure 4-42. Warhead Details (Discrete Rod Warhead)

nation is virtually impossible without carrying out specific tests. Generally, Composition B, H-6, 80/20 Tritonal and Composition C-3 are used.

The initial velocity imparted to the rods is a function of the ratio of the weight of charge to the weight of the rods, that is c/m. The c/m required to impart rod velocities in the neighborhood of 2000 to 3000 feet per second is of the order of 1/4. A satisfactory c/m to produce a given initial velocity can only be determined by test since it depends upon rod breakup, rod length and diameter, shape of explosive charge, end plate effects, shape of rod bundle and other interrelated variables which have not been investigated experimentally.

Warhead Details The individual rods are packaged in the warhead around the explosive charge by lightly tack-welding them together at their ends or by lightly welding or brazing them to a supporting tube at their ends. More than one layer of rods is generally used. A liner or filler is usually employed to separate the explosive cavity from the rods. See Figure 4-42. The shape of this filler is often varied in an effort to minimize rod breakup. The explosive cavity usually includes an inner liner. End plates are added to confine the charge. Provision for loading the explosive is usually made in the end plates. The booster should be centrally located to provide an even distribution of initial rod velocity. Installation and handling fittings must be provided as required by the missile system designer.

Summary of Fuzing Requirements The fuze designer needs design information to design a fuze which is compatible with the missile system and the warhead. He will have access to the same missile system data as did the warhead designer. In addition to this, the fuze designer will need the following information relating specifically to the warhead.
 (1) Type of Explosive Used
 (2) Drawing of Warhead
 (3) Rod Length and Cross Section
 (4) Initial Rod Velocity
 (5) Missile Velocity

Summary of Design Data At the conclusion of the design procedure, a summary of engineering data relating to the warhead should be prepared. This should include the following items:
 (1) Total Weight
 (2) Design and Installation Drawings
 (3) Explosive
   (a) Material
   (b) Weight
 (4) Charge to Metal Ratio
 (5) Rods
   (a) Length and Cross Section
   (b) Number
   (c) Material
   (d) Weight
   (e) Initial Rod Velocity (Static)
 (6) Location of Center of Gravity
 (7) Method of Mounting

4-4.3. References

4-4.a Report NMIMT/RDD/T-821, "Rod Design", M. L. Kempton and C. R. Cassity, February, 1952, Research and Development Division, New Mexico Institute of Mining and Technology, ASTIA AD 47605.

4-4.4. Bibliography

(1) NPG Report No. 1106, April 1953, "Terminal Ballistics of Rodlike Fragments", U.S. Naval Proving Ground, Dahlgren, Virginia.

4-5. CONTINUOUS ROD WARHEADS

4-5.1. Detail Design Steps

Step
No.   Detail Design Step
 1. Determine Rod Cross Sectional Dimensions
 2. Determine Dimensions of Rod Bundle
 3. Select Type and Amount of Explosive

4. Design Explosive Cavity
5. Design Warhead Details
6. Prepare Summary of Fuze Data
7. Prepare Summary of Design Data

The exact order of the design procedure may vary depending upon the viewpoint of the designer and, even more, on the military requirements which often fix certain parameters in advance.

4-5.2. Detail Design Data

Rod Cross Sectional Dimensions The cross sectional dimensions of the rods are established by considering target damage requirements. Tests have been conducted at the New Mexico Institute of Mining to determine the various parameters which affect damage to aerial targets. From these tests it has been determined that rods with square cross sections 3/16 and 1/4 inches on a side striking at more than 3500 and 3000 feet per second, respectively, are lethal if they strike a vulnerable portion of the target. Thus, as a first approximation, the cross sectional dimensions of the rods can be selected between these two narrow limits.

Dimensions of Rod Bundle The length of the rods depends primarily upon packaging limitations. For a given weight and space allocated to the warhead by the missile designer, the warhead should be designed to provide the largest possible expanded hoop radius. This is desirable because the lethality drops rapidly if the rods do not strike the target until after the hoop has expanded to the point where it is no longer continuous (see Figure 4-43). Therefore, the greater the fully expanded radius, the greater the allowable guidance error for high lethality.

The fully expanded radius is obviously a function of the summation of the lengths of the individual rods. Approximately 65% of the total warhead weight may be allotted to the steel rods, this percentage being typical of most successful continuous rod warhead designs. Knowing the total weight of all of the rods, their individual cross sectional dimensions and the density of steel, the total length of rod material may be computed. It then remains to decide on the length of the individual rods, which establishes the length of the rod bundle.

Since the cross sectional dimensions are fixed, the number of rods chosen fixes the diameter of the bundle. Certain limits are imposed on the rod bundle length and diameter and L/D, length to diameter ratio. Obviously, the length and diameter cannot exceed the warhead compartment dimensions. The length should be as long as possible so as to minimize the hoop radius lost due to the welded end portions of adjacent rods.

However, the length should be limited to between 2 to 3 times the diameter of the bundle. At L/D values in excess of 3, difficulty will be experienced in designing the explosive charge so as not to cause bending and distortion of the individual rods upon detonation. At low values of L/D, the expanded hoop radius will be shortened since the length of the rod used for welding cannot contribute to the hoop circumference. Values of L/D as low as 1 have been used. The actual expanded hoop radius is approximately 70% to 85% of the theoretical radius, based on the summation of the lengths of the rods. The 70% figure applies to bundles with low values of L/D and also allows for imperfect expansion of the hoop.

Explosive Charge The type and amount of explosive charge used is based on the initial rod velocity required to provide a striking velocity which will kill the target. The striking velocity depends upon the vector sum of the missile and initial rod velocities and the loss in velocity due to air resistance as the hoop expands in its flight toward the target. A striking velocity of 3000 feet per second for 1/4 x 1/4 inch rods and 3500 feet per second for 3/16 x 3/16 inch rods is considered a lower limit. (See Reference 4-5.b). Rod velocity as a function of radial distance from the detonation point for 1/4 x 1/4 inch and 3/16 x 3/16 inch rods has been obtained experimentally in Reference 4-5.a. This information is presented in Figure 4-44.

The ratio of the explosive charge to the weight of the rod material (c/m ratio) is the most significant parameter affecting initial rod

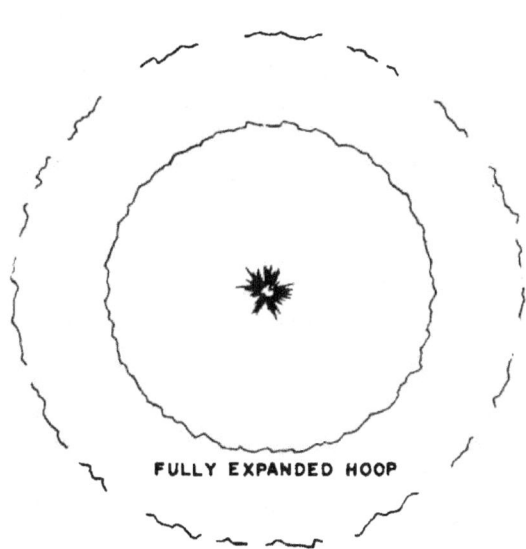

Figure 4-43. Expansion of Rod Hoop

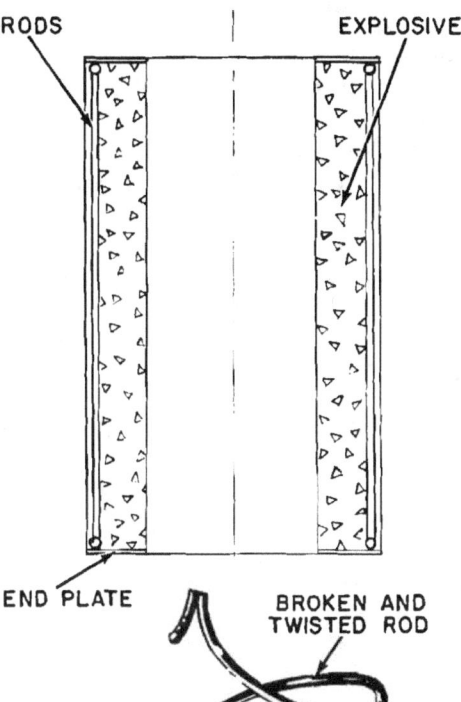

Figure 4-45. A Warhead Design Producing Tangled and Broken Rods

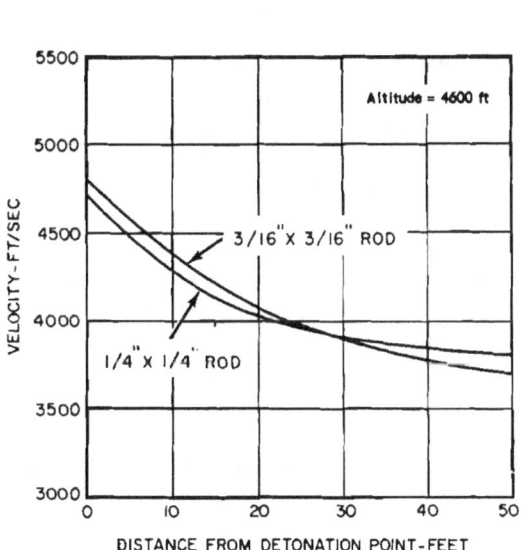

Figure 4-44. Rod Velocity Variation

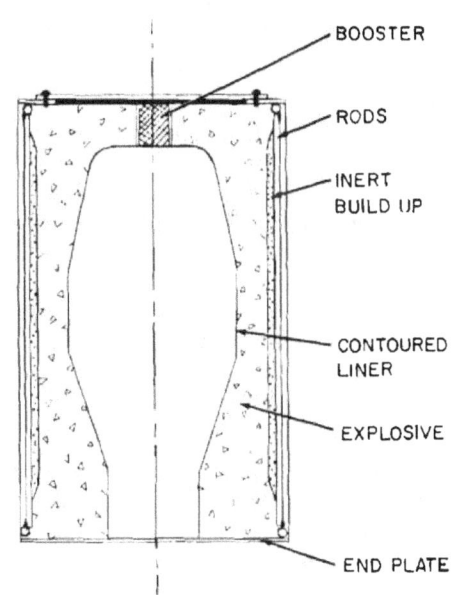

Figure 4-46. Contoured Liner and Inert Build Up

velocity. Hollow central cavity warheads employing 3/16 x 3/16 to 1/4 x 1/4 inch rods and a c/m of between .60 and .75 will produce initial velocities of approximately 5000 feet per second. The exact value of the c/m required will depend upon the amount of charge confinement afforded by the inner wall of the explosive cavity, the geometry of the explosive cavity (discussed in the following subchapter) and upon the type of explosive used. Composition C-4, B and H-6 are suitable types for this application. (See Appendix.)

Explosive Cavity The explosive cavity usually has a hollow center. The cavity must be designed so that detonation of the explosive charge imparts a constant velocity to the rod along its entire length, thereby accelerating the rod without excessively bending or distorting it. The use of an explosive loading with a constant cross section along the entire length of the rod bundle is always accompanied by rod tangling and excessive rod breakage. These undesirable conditions are caused by the higher velocities imparted to the central portions of the rod bundle as compared to the ends. Figure 4-45 shows a warhead with an explosive loading of uniform cross section and a broken and twisted rod typical for this design. Contoured liners and the use of inert material placed along the length of the rod bundle have proved to be very effective in eliminating tangling and rod breakage due to differential rod velocities. Figures 4-46 and 4-47 show contoured liner and inert build up geometries used in successful designs. These figures should be used as the basis of designs for new warheads. Abrupt angular changes in the contours of the liners and in the inert build up or fillers are to be avoided. A number of materials can be used for the inert build up. Paraplex polyester resin has been used extensively as an inert compound because it is tough, hard, easily formed by casting or machining, and is infusible at moderate temperatures. Plaster of Paris has also been used successfully.

The placement of the booster charge has an important effect on rod breakup at detonation. When the booster is placed in the annular explosive ring of a hollow warhead, the rods on the side opposite the booster break during detonation. The general practice for hollow warheads is to place the booster on the warhead axis either symmetrically within the annulus or in the center of a cylindrical plate of explosive located across one end of the warhead as shown in Figure 4-46. In the case of solid warheads the booster should be placed on the warhead axis as near midway between the ends of the rod bundle as possible. This is because the portion of the rod bundle which surrounds the booster will receive a lower radial velocity than more distant portions of the rod bundle. Extension of the explosive charge beyond the end of the rod bundle is sometimes used in warheads of small diameter to overcome rod velocity loss in the vicinity of the booster. The booster is then placed beyond the end of the rod bundle in the explosive extension.

Scabbing as exemplified by Figure 4-48, and surface damage to rods during detonation are deleterious effects usually attributed to the numerous small gaps and openings which exist between adjacent rods of the rod bundle in direct contact with the explosive loading of the warhead. The scabbing problem may be overcome by filling these gaps which open in the rod bundle and which are in contact with the explosive loading. Experiments at the New Mexico Institute of Mining and Technology show that commercial white lead applied in a thin coating about .005 to .020 inches in thickness is a nearly ideal solution to the scabbing problem. Plastic laminacs have also been used for this purpose. Masking tape or other material must be used to cover the coating to prevent mixing of the white lead and the explosive during loading (see Reference 4-5.b). Navy standard practice is to coat all metal surfaces of the explosive cavity with cavity hot melt (Code 280-3110-0) before casting the explosive. For continuous rod warheads the recommended hot melt thickness is 1/64 to 1/32 inch.

A series of successful continuous rod warhead designs which have been tested are presented in Figures 4-49a through 4-49l. These

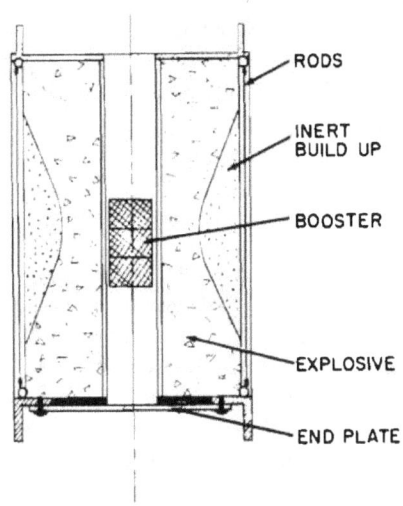

Figure 4-47.
Contoured Inert Build Up

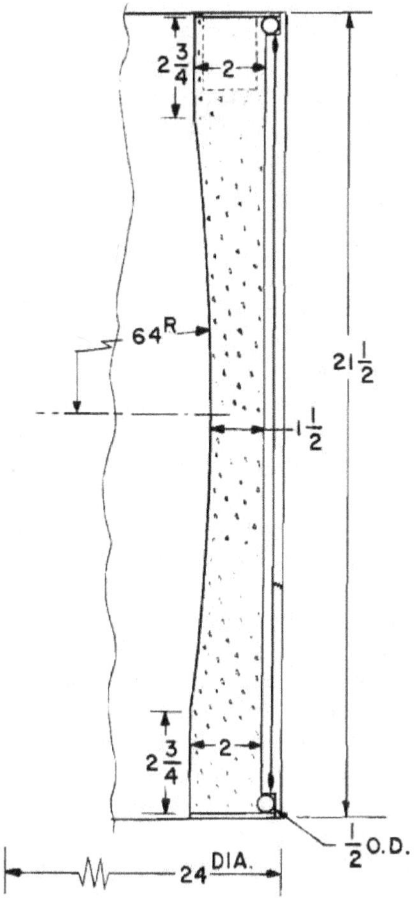

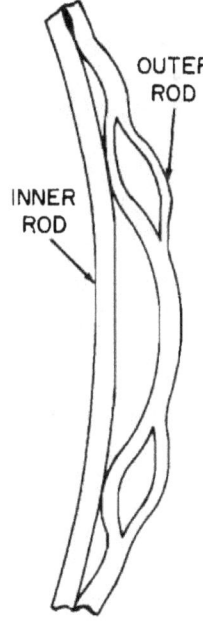

Figure 4-48.
Rod Scobbing

Design Details

Gross weight	414 lb
Explosive wt. (C-4)	129 lb
Rod size	1/4 x 1/4 in.
End plates	1/4 in.
Booster	Tetryl

Warhead Performance

Initial rod vel.	4500 ft/sec.
Rod velocity at 90 ft radius	3000 ft/sec.
Max. opening R.	120 ft

Figure 4-49a. Continuous Rod Warhead Designs

include a sketch of the complete warhead (with lengths in inches), data relating to the warhead components and information on the performance of the warhead. They were taken from Reference 4-5.a and are included here to guide the warhead designer particularly in regard to the design of the explosive cavity.

Warhead Details All continuous rod warheads must be provided with end plates which serve to hold the warhead together, to contain the explosive at the ends of the warhead and in some cases to serve as structural members. Good containment also helps to decrease rod bending and breakage. Depending on warhead size and design, the thickness of the end plates will vary from about 0.125 to 0.375 inch. In annular warheads, weight can be saved by eliminating the central portion of the plate. Additional weight and simplicity can often be achieved by combining end plates and warhead-to-missile attachment fittings. Rod pairs are generally joined at opposite ends of arc welds or resistance welds so that the rod blanket consists of a double layer of rods. The outer rod usually contains a small tab at each end for attaching the rod blanket by welding the tabs to the end plates. The inner rods are grooved at each end to accept 360° cutoff tubes. The so-called cutoff tubes are located at both ends of the warhead between the ends of the rod bundle and the end plates, as shown in Figure 4-49a. The function of cutoff tubes is to release the rods from the end plates during detonation by collapsing and forming a modified shaped charge effect to sever the end tabs from the rods. Fuzing and booster provisions must be made as well as provisions for handling and installing the warhead in the missile com partment.

It is to be noted that facilities at the New Mexico Institute of Mining and Technology are utilized for most rod warhead testing.

Summary of Fuzing Requirements The fuze designer needs design information to design a fuze that is compatible with the missile system and warhead. He will have access to the same missile system data as did the warhead designer. In addition to this, the fuze designer will need the following information relating specifically to the warhead:
    (1) Type of Explosive Used
    (2) Drawing of Warhead
    (3) Rod Length and Cross Section
    (4) Expanded Rod Diameter
    (5) Initial Rod Velocity (Static)

Summary of Design Data At the conclusion of the design procedure, a summary of engineering data relating to the warhead should be prepared to summarize the design. This should include the following items:
    (1) Total Weight
    (2) Design and Installation Drawings
    (3) Explosive
        (a) Material
        (b) Weight
    (4) Charge-to-Metal Ratio
    (5) Rods
        (a) Length and Cross Section
        (b) Expanded Rod Diameter
        (c) Material
        (d) Weight
        (e) Initial Rod Velocity (Static)
    (6) Location of Center of Gravity
    (7) Method of Mounting

4-5.3. References

4-5.a "Guide to the Design of the Continuous Rod Warhead", M. L. Kempton, Report No. NMIMT/RDD/T-922, New Mexico Institute of Mining and Technology, ASTIA AD 90709, March 1956.

4-5.b "Effectiveness of Talos Continuous Rod Warhead (Revised)", P. M. Whitman, APL/JHU CF-2111A, December 1953.

4-5.4. Bibliography

(1) "Supplement to Effectiveness of Talos Rod Warhead", P. M. Whitman, APL/JHU CF-2111B, May 1954.

(2) "Comparison of Different Warheads for the Hawk Missile Against the IL 28-2 and

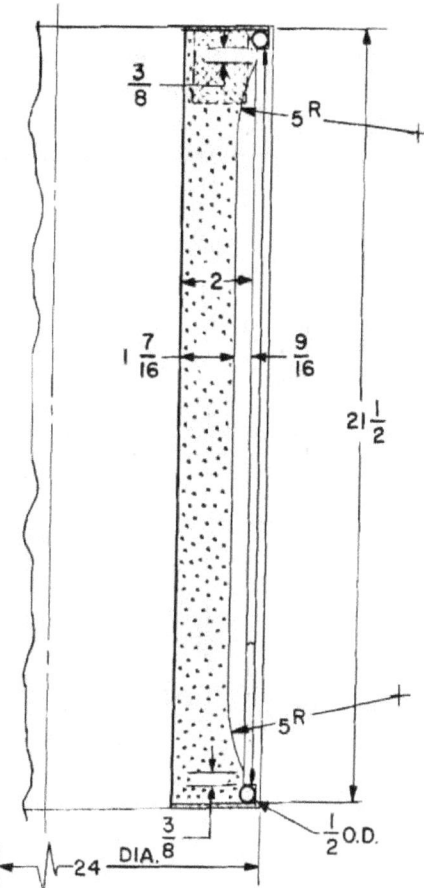

Design Details

  Gross weight   394 lb
  Explosive wt. (C-4) 109 lb
  Rod size   1/4 x 1/4 in.
  End plates   1/8 in.
  Booster   Tetryl

Warhead Performance

  Initial rod vel.   4500 ft/sec.
  Rod vel. at
    90 ft radius   3000 ft/sec.
  Max. opening
    radius   120 ft

Figure 4-49b. Continuous Rod Warhead Designs

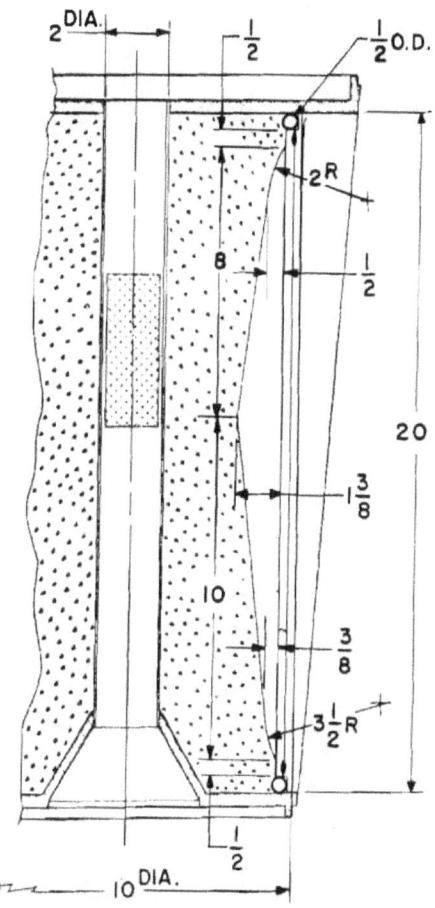

Design Details

  Gross weight   171 lb
  Explosive wt. (C-4) 41 lb
  Rod size   3/16 x 1/4 in.
  End plates   3/8 in.
  Booster   Tetryl and C-3

Warhead Performance

  Initial rod vel.   4700 ft/sec.
  Rod velocity at
    71 ft radius   3200 ft/sec.
  Max. opening
    radius   71 ft

Figure 4-49c. Continuous Rod Warhead Designs

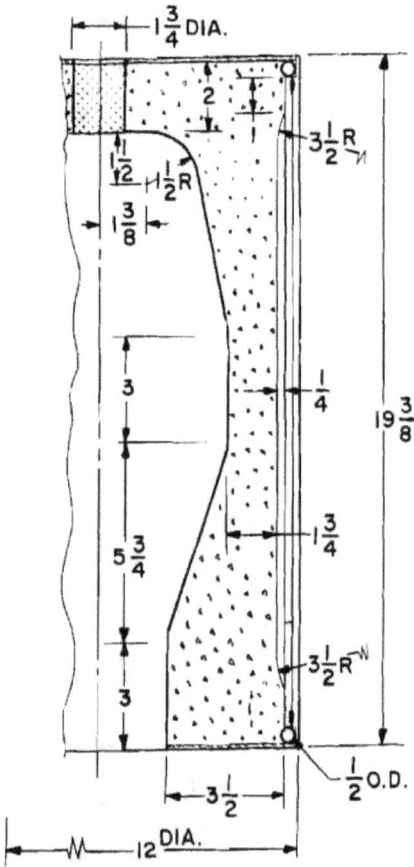
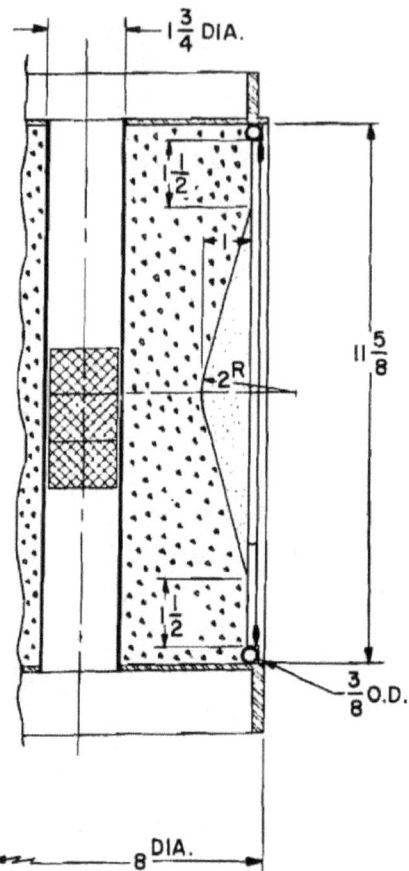

Design Details

Gross weight	178 lb
Explosive wt. (C-4)	59 lb
Rod size	1/4 x 1/4 in. or 3/16 x 1/4 in.
End plates	1/8 in.
Booster	Tetryl

Warhead Performance

Initial rod vel.   4700 fps
1/4 x 1/4 in. rods:
    Max. opening R.   63 ft
    Rod vel. at 63 ft R. 3600 fps
3/16 x 1/4 in. rods:
    Max. opening R.   84 ft
    Rod vel. at 84 ft R. 3200 fps

Figure 4-49d. Continuous Rod Warhead Designs

Design Details

Gross weight	64 lb
Explosive wt. (C-4)	18 lb
Rod size	3/16 x 3/16 in.
End plates	fore 3/16 in. aft 1/8 in.
Booster	Tetryl

Warhead Performance

Initial rod vel.	4600 fps
Rod vel. at 32 ft R.	4000 fps
Max. opening R.	32 ft

Figure 4-49e. Continuous Rod Warhead Designs

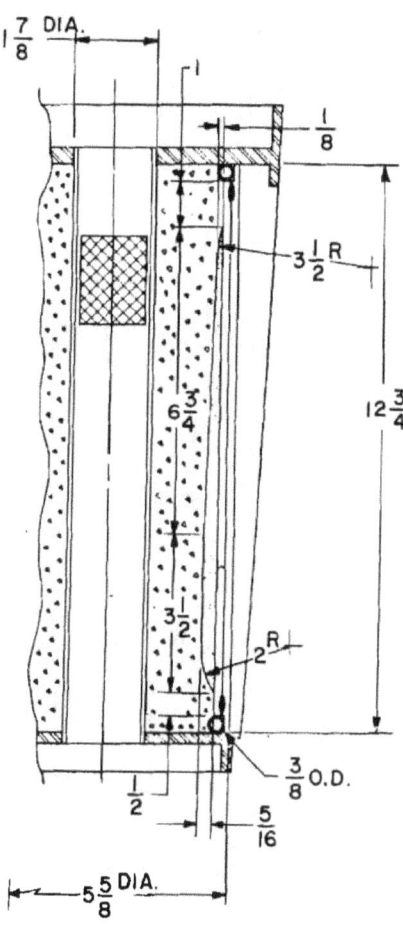

Design Details

    Gross weight   45 lb
    Explosive wt. (C-4)   9 lb
    Rod size   3/16 x 3/16 in.
    End plates   fore 1/4 in.
                      aft 3/8 in.
    Booster   Tetryl

Warhead Performance

    Initial rod vel.   4200 fps
    Rod vel. at 24 ft R.   3600 fps
    Max. opening R.   24 ft

Figure 4-49f. Continuous Rod Warhead Designs

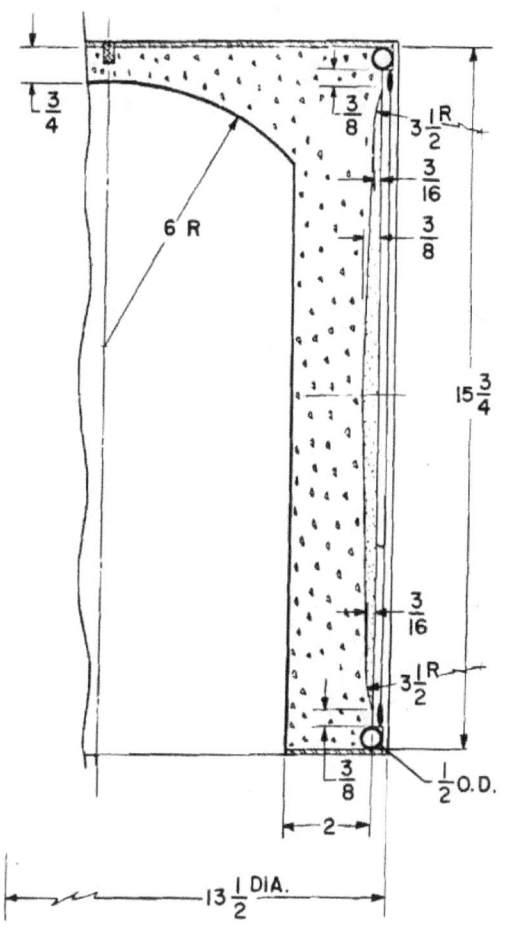

Design Details

    Gross weight   137 lb
    Explosive wt. (C-4)   53 lb
    Rod size   3/16 x 1/4 in.
    End plates   1/8 in.
    Booster   P-11

Warhead Performance

    Initial rod vel.   5000 fps
    Rod vel. at 57 ft R.   3700 fps
    Max. opening R.   57 ft

Figure 4-49g. Continuous Rod Warhead Designs

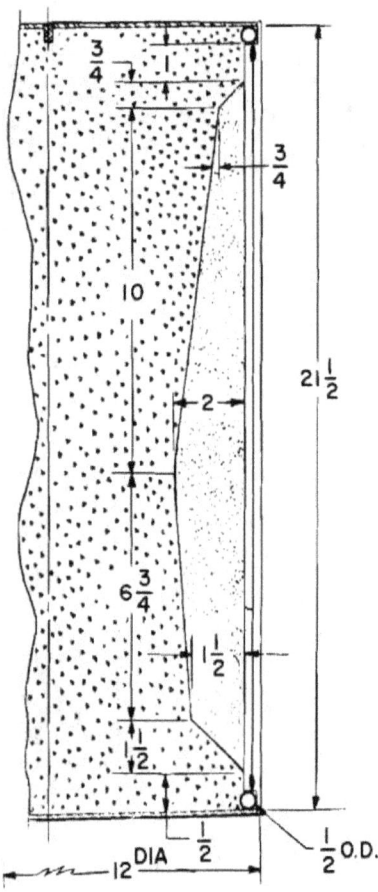
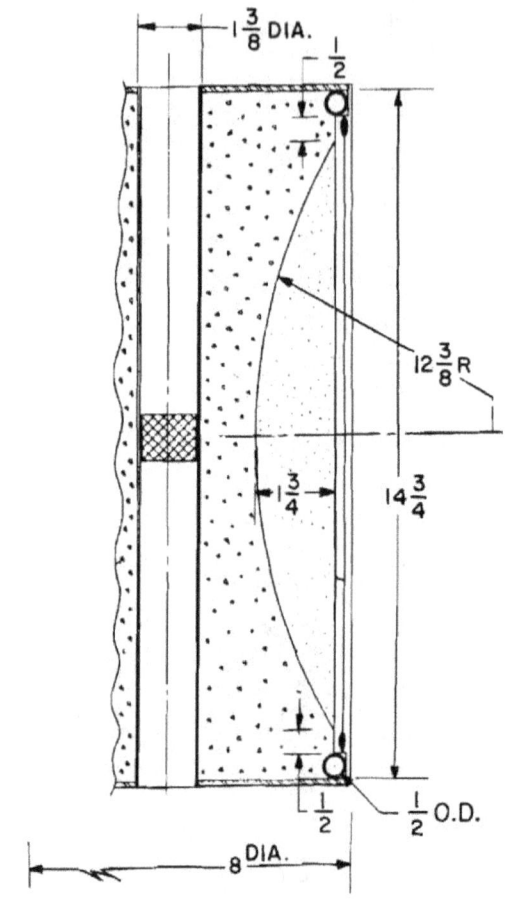

Design Details

Gross weight	210 lb
Explosive wt. (C-4)	65 lb
Rod size	1/4 x 1/4 in.
End plates	fore 1/8 in.
	aft 1/4 in.
Booster	P-11

Warhead Performance

Initial rod vel.	5000 fps
Rod vel. at 71 ft R.	3500 fps
Max. opening R.	71 ft

Figure 4-49h. Continuous Rod Warhead Designs

Design Details

Gross weight (C-4)	60 lb
Explosive weight (C-4)	15 lb
Gross weight (B)	64 lb
Explosive weight (B)	19 lb
Rod size	3/16 x 3/16 in.
End plates	3/32 in.
Booster	MK 44 Aux. Det.

Warhead Performance

Initial rod vel.	4500 fps
Rod vel. at 41 ft R.	3700 fps
Max. opening R.	41 ft

Figure 4-49i. Continuous Rod Warhead Designs

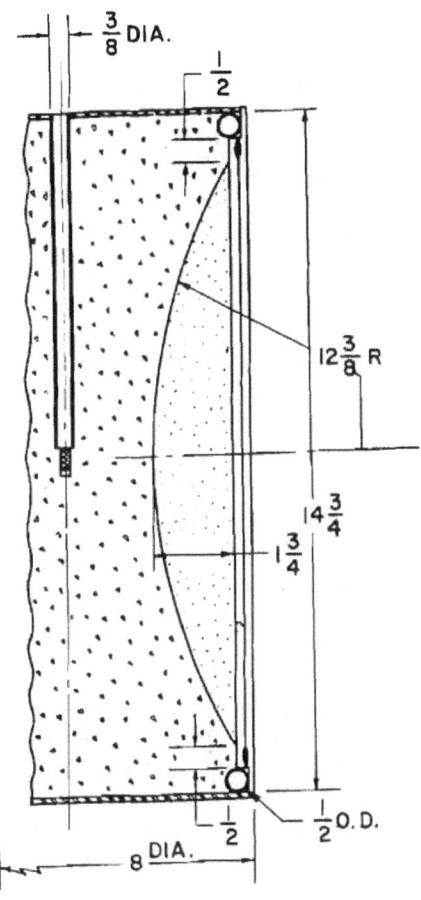
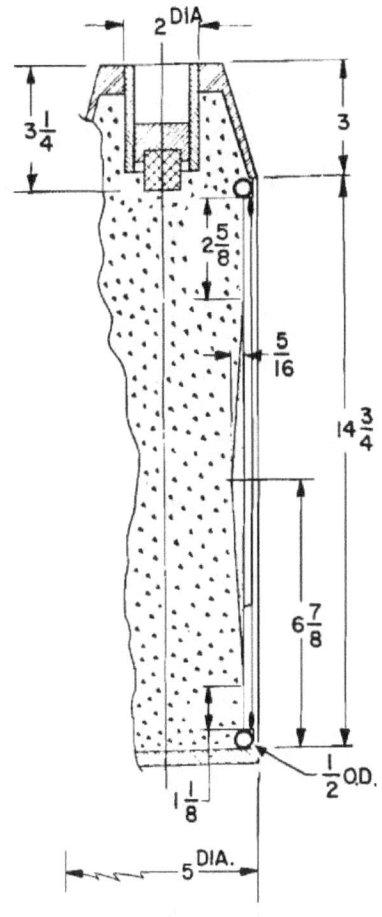

Design Details

    Gross wt.   57 lb
    Explosive wt. (C-4)   17 lb
    Rod size   3/16 x 3/16 in.
    End plates   3/32 in.
    Booster   P-11

Warhead Performance

    Initial rod vel.   4500 fps
    Rod vel. at 41 ft R.   3700 fps
    Max. opening R.   41 ft

Design Details

    Gross weight   40 lb
    Explosive wt. (C-4)   11 lb
    Rod size   3/16 x 3/16 in.
    End plates   aft 3/8 in.
    Booster   MK 44 Aux. Det.

Warhead Performance

    Initial rod vel.   5000 fps
    Rod vel. at 23 ft R.   4500 fps
    Max. opening R.   23 ft

Figure 4-49j. Continuous Rod Warhead Designs

Figure 4-49k. Continuous Rod Warhead Designs

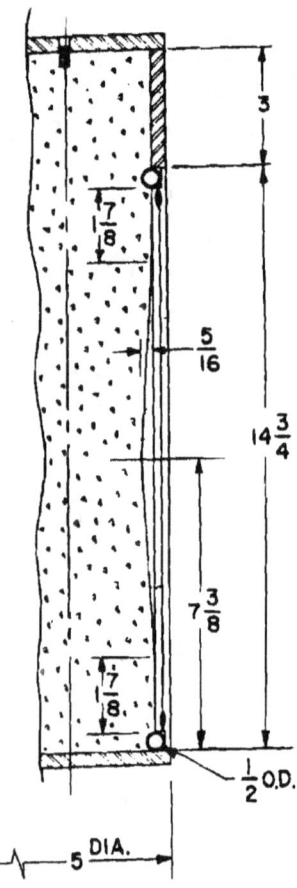

Design Details

Gross weight	42 lb
Explosive wt. (C-4)	12 lb
Rod size	3/16 x 3/16 in.
End plates	3/8 in.
Booster	P-11

Warhead Performance

Initial rod vel.	5000 fps
Rod vel. at 23 ft R.	4500 fps
Max. opening R.	23 ft

Figure 4-49I. Continuous Rod Warhead Designs

B-29 Aircraft", W. Taylor Putney, Ed S. Smith and G. Trevor Williams, BRL Memo. Report 905, July 1955.

(3) "Guidebook to Antiaircraft Guided Missile Warheads, Coordination of Fuze and Warhead", Philip M. Whitman, APL/JHU CF-2419, undated (about 1955).

(4) "Guidebook to Antiaircraft Guided Missile Warheads, Effectiveness of Rod Warheads", Philip M. Whitman, APL/JHU CF-2486, February 1956.

4-6. CLUSTER WARHEADS

4-6.1. Detail Design Steps

Step No.	Detail Design Step
1.	Estimate the Optimum Pattern
2.	Select the Type of Submissile
3.	Select the Ejection Method
4.	Determine the Maximum Number of Submissiles
5.	Design in Detail the Ejection System
6.	Design in Detail the Submissile
7.	Design in Detail the Support Structure
8.	Design in Detail the Retention System
9.	Design in Detail the Obstruction Removal Device
10.	Prepare Summary of Fuzing Design Data
11.	Prepare Summary of Design Data

The exact order of the design procedure may vary depending upon the viewpoint of the designer and, even more, on the military requirements which often fix certain parameters in advance.

4-6.2. Detail Design Data

Optimum Pattern Cluster warheads are ideally designed to eject a series of submissiles from the warhead compartment in a pattern such that the probability of one or more submissiles striking the target is a maximum. In one design of antiaircraft warheads, for example, the submissiles are arranged in rows around the periphery of the warhead compartment and are ejected radially. Upon ejection the submissiles form a radially expanding circular pattern,

Table 4-8 Summary Chart

Designation	Total Weight pounds	Metal Wt. (Less Fins and Fuze) pounds	HE Weight pounds	Diameter inches	Length inches	General Shape	Results of Penetration Tests
T-214 With Windshield	4.20	2.30	1.35	2.00	14.2	Cylinder with ogive nose	Good - 80° obliquity 1/2 in. 75 S-T6 Penetration Good with 30° yaw
T-214 Without Windshield	4.20	2.25	1.35	2.00	12.5	Cylinder with slight ogive and flat nose	" "
Gimlet	4.00	2.25	1.20	2.00	11.0	Cylinder with ogive nose	Break up at 80° obliquity 1/2 in. 75 S-T6
Sprite Sparrow	4.00	1.60	1.40	Keystone	11.5	Keystone cross section; tapered longitudinally with folding fins and flat nose	Break up at 70° obliquity 1/4 in. 75 S-T6
Sprite Talos	5.00	2.60	1.40	Keystone	11.5	Keystone cross section with folding fins, flat nose	
Edgewood E91 Types				2.82 to 3.44	9.88 to 15.00	Tear drop with fixed fins	Break up on ground impact due to light case, therefore poor target plate penetration likely
Dart	1.33		.25	.98	19.0	Cylinder with ogive and fins	Good - 70° obliquity 3/8 in. 75 S-T6
Dart	1.70		.36	.98	22.4	" "	
Dart	3.90		1.00	1.58	23.5	" "	
Aeroflak	5.6		2.75	2.80	5.5	Short cylinder with ogive	Good - 70° obliquity 3/8 in. 75 S-T6

generally with the submissiles in one or more rings. The pattern is moving toward the target at missile speed and expanding radially at ejection speed. The initial velocity of individual submissiles is the vector sum of the missile and ejection velocities. (See Figure 4-50.) The striking velocity is the vector difference between the final actual submissile velocity and the target velocity. The radius of the pattern at any instant is a function of the time of flight after detonation and the average radial velocity. It is desirable to keep the flight time to a minimum to reduce submissile slow-down due to aerodynamic drag forces, to reduce the effect of gravitational forces and to reduce the effectiveness of evasive action of the target. Cluster warheads are usually designed to produce a circular submissile pattern whose radius is slightly greater than the standard error of guidance of the missile system at the instant the plane of the submissile circle reaches the target. Flight times of from .3 to .75 seconds with ejection velocities from 200 to 400 feet per second are consistent with guidance errors (or pattern radii) of 40 to 200 feet.

Types of Submissiles The submissiles used in the cluster warhead are of two general types: stabilized and unstabilized. Typical examples of each type are shown in Figures 4-51 through 4-56. The unstabilized type requires the use of an "all-ways fuze", one which insures detonation of the submissile regardless of its orientation when striking the target. It is suggested that the fuze designer be consulted to determine the adaptability of this type of fuze to the particular warhead design under consideration. If the "all-ways fuze" can be utilized, the optimum type of submissile is generally the unstabilized one. The unstabilized submissile has certain advantages over the stabilized type. It is easier to manufacture and assemble, and requires less volume because no stabilization devices are necessary. Therefore a larger number of submissiles can be incorporated in a given warhead volume. On the debit side, the unstabilized submissile, in addition to requiring the aforementioned "all-ways fuze", is subjected to more severe drag forces during its flight to the target.

The stabilized submissile is used in the event that an "all-ways fuze" is not applicable. This enables the designer to specify a lighter and simpler fuze which detonates the submissile after penetration of the target. This in turn requires a smaller amount of explosive since the blast is internal. However, this type of submissile requires a structural nose, usually made of steel and a stabilization mechanism, each of which tends to increase the submissile weight.

In present designs, stabilization is accomplished by use of a drag tube, drag plate, drag chute, fixed fins or collapsible fins. The selection and design of these mechanisms is discussed later. The stabilization mechanism must be stowed with the submissile which obviously increases the weight and volume and, even though the charge is less, the net effect for a fixed total warhead weight is fewer submissiles as compared to the unstabilized type. Packaging of stabilized submissiles becomes difficult and, if a stabilizer release mechanism is included, it must be of a rugged design to withstand the ejection accelerations and aerodynamic forces.

Ejection Methods The function of the ejection system is to impart a velocity to the submissile in a direction normal to the missile axis. Current systems utilize gas pressure generated by the burning of a suitable propellant. Gas pressure systems may also be divided into two general categories: the gun tube and the blast type. Both types have been successfully used in developmental missiles.

The major difference between the two ejection systems is as follows. In the blast type the explosion emanates from a central source or chamber and the resulting pressure is either directed through ports or openings to act on the submissile or acts directly on the submissiles. In the "integral gun type" each submissile is fired by a charge acting in an individual gun chamber. The blast type will therefore have the advantages of a minimum space requirement and a simplified firing mechanism since individual submissile firing mechanisms are eliminated. On the other hand, use of the integral gun type results in a more uniform submissile pattern. An alternate means of submissile ejection is to depend upon aerodynamic forces to launch the submissiles from the missile after the missile skin has been severed.

Figures 4-57 through 4-62 illustrate typical ejection systems used on current warheads. A brief description of the illustrated systems follows.

Gun Types The gun ejection system shown in Figure 4-57 is of the integral ignition type. The ejector guns are separate units screwed into the backup ring, which serves as a structural member. They are actuated electrically and each submissile has its own primer and dispersal charge. The submissiles are placed over the ejector gun tubes, and are packed in rings around the warhead. The burning of the propellant in the gun generates pressure that acts on the ejector tube which is part of the submissile, and imparts a force to the submissiles.

The gun ejection system shown in Figure 4-58 is an example of the central ignition type. This type is similar to the integral ignition system except that the ignition of the gun charge is effected by firing a central source of powder instead of individual igniters. The ring containing the powder acts as a structural member and absorbs the ejection forces. The powder charge is simultaneously fired by several primers located around the chamber.

Another variation of the gun type ejection system is known as the piston type, shown in Figure 4-59. In this system a dispersal gun is used which consists of two gun chambers connected by a steel igniter tube. The gun is fired electrically at one end. When the propellant is ignited, the hot gases expand in the chamber and actuate pistons which eject the submissiles. Ignition of the charge in the front chamber is caused by hot gases from the rear chamber flowing through the igniter tube. Each piston is equipped with an "O" ring to minimize gas leakage, thus giving rise to higher pressures and resulting in high ejection velocities. In this design, the gun chambers may also be utilized as a structural part of the parent missile (Reference 4-6.d)

Blast Types Figure 4-60 shows a blast system applied to one submissile. The same method can be adapted to many submissiles in a complete warhead. The source of energy for ejection is a propellant contained within the confines of a pressure tube, known as a backup tube. A liner of predetermined breaking strength separates the propellant from the submissiles. This liner fits the inside diameter of the backup tube snugly.

The backup tube contains three orifices per submissile. Upon ignition the propellant gas expands, thereby creating pressure against the inner surface of the liner. This expansion causes failure of the liner and permits the gases to impinge upon the base of the submissiles. The force so created causes the submissiles to be ejected laterally from the backup structure (Reference 4-6.b).

The device shown in Figure 4-61 is of the blast ejection segmented chamber type. It uses the segments that surround the propellant cavity as sabots for the submissiles. This provides good control of both the submissile pattern and individual submissile velocity. At present little is known concerning the ejection transients inherent in this method. Selection of this type would call for extensive tests to determine its feasibility (Reference 4-6.c).

The method illustrated in Figure 4-62 appears to be one of the better methods devised to date in that a convoluted expanding liner around a ported chamber acts as a gas seal during the initial phase of ejection. This liner also offers some protection to the fins of stabilized type submissiles from the high pressure gases, and distributes the ejection forces more equally over the submissile body. Another advantage is that the ported chamber can be used as a structural part of the parent missile (Reference 4-6.c).

Number of Submissiles When designing an optimum cluster warhead without regard to missile

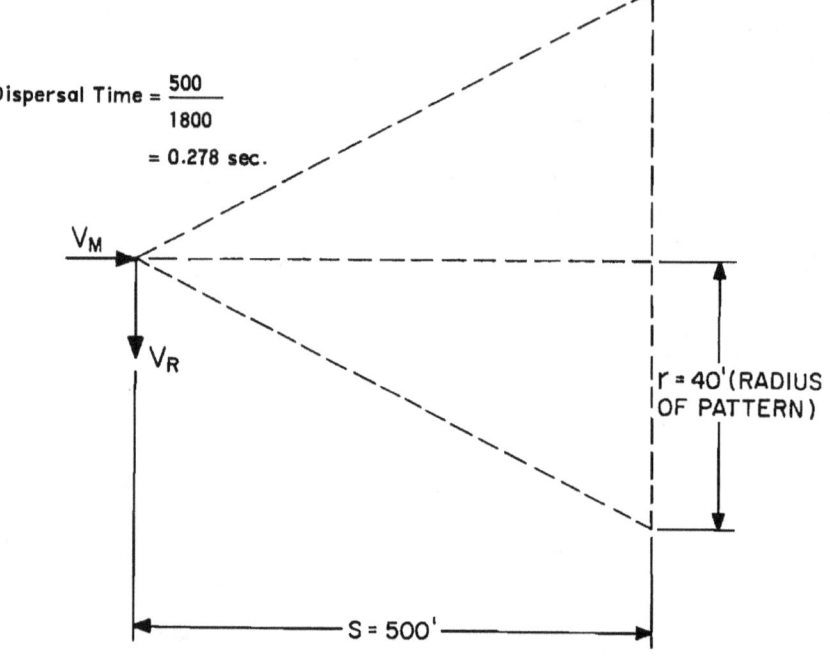

Average velocities are proportional to distances

$$\frac{V_R}{V_m} = \frac{r}{s}$$

$$\frac{V_R}{1800} = \frac{40}{500}$$

$$V_R = 144 \text{ ft/sec.}$$

Dispersal Time $= \frac{500}{1800}$
$= 0.278$ sec.

Figure 4-50. Resolution of Velocities

space and weight limitations, the number of submissiles is determined by the number needed on a kill probability basis. Hence, one assumes the mission of the missile is to provide a specified kill probability, and generally starts with the necessary number of submissiles on the conditional probability that a hit is a kill. However, if the warhead is being designed for an existing missile, the following different approach can be used.

The maximum number of submissiles that can be installed in a cluster warhead is a function of the space and weight available for the warhead and backup structure, and the size and weight of the individual submissiles. For some configurations, it is evident that a number of lightweight submissiles may fill up the available warhead space, but the warhead will be lighter than the allowed weight. For other configurations, the allotted space in the warhead will be so great that the weight allowed will be reached, but the warhead volume will not be completely filled. However, each submissile must contain enough explosive to be effective.

In the case of the unstabilized submissile designed for detonation external to the target, the majority of antiaircraft warheads built to date have used from two to three pounds of HBX explosive per submissile, depending on the target. The two pound charge may be assumed satisfactory for a small target (such as fighter aircraft), and the three pound charge may be used when the target is larger (such as a bomber). In the case of either stabilized submissiles or unstabilized submissiles designed for target penetration, most antiaircraft warheads built to date have utilized from one to two pounds of HBX, again depending on the target. These lower values are effective due to the fact that the submissiles that penetrate the target require less explosive to produce equal damage. A more complete discussion of blast effects is given in subchapter 4-2. After assuming the weight of the explosive per submissile, the designer can determine the number of submissiles that can be utilized.

To properly determine the number of submissiles, one must calculate the allowable number on both a weight and volume basis. The values of the number of submissiles from the two sets of calculations should be compared, and the smaller integral number chosen as the maximum number of submissiles which can be utilized, since this number satisfied both the weight and volume requirements.

No set form can be presented to assist the designer in the determination of the maximum number of submissiles that can be obtained. The designer must work around his set values, such as weight of warhead, center of gravity of warhead, and available space in the parent missile for warhead structure and submissiles. From this, the problem resolves into one of geometrical relationships and, once a shape is decided upon, the number of submissiles can be determined.

The approximate number of submissiles determined should be checked at the completion of the warhead design by laying out the entire unit. The location of the warhead may be such that it lies in the ogive or tapering section of the missile, with a difference of several inches or more between the fore and aft diameters of the section. In this case, to obtain the maximum number of submissiles, they must be arranged in rows, with the diameters increasing progressively by steps. A number of trial solutions must be investigated to obtain the optimum number of submissiles, bearing in mind that the design total weight of the warhead must not be exceeded and that the location of the center of gravity of the warhead must be adhered to. If the total length of the warhead is such that six or more rows of submissiles will be practical, consideration should be given to varying the ejection velocities progressively in the rows, thus producing a pattern of submissiles ejected in space, consisting of concentric circles of different diameter as shown in Figure 4-63. This is generally preferable to having a cylindrical pattern of submissiles spaced very close together. However, a noteworthy exception is a case in which an analysis similar to that used in the design of continuous rod warheads shows that the expected number of hits on a target would be greater with a single expanding circle of submissiles. This can result in cases where the average bias due to air drag on the submissiles brings the single circle closer to the center of area of the target than if the submissiles are distributed over a number of circles.

Because of practical considerations, the number of steps of the external and internal diameters of the submissiles should be kept to a minimum. The lengths (fore and aft) of the cases should also be a minimum value, and should be kept equal for all submissiles.

Where there are several similar rows of submissiles all ejected at the same velocity, it is advisable to stagger the angular location of the submissiles in the successive rows, so as to provide the most even distribution of the ejected submissiles around the circles in space.

Design of Ejection System

Gun Tube Method In the gun tube method of ejection, the submissiles are ejected by the burning of a black powder propellant in a steel gun tube, mounted radially on a fixed central support structure, and projecting into a closed steel ejection tube (Figure 4-64) which is an integral part of the submissile as shown in Figure 4-65.

The radial ejection velocity depends upon the pressure generated by the propelling charge, the area of the ejection tube bore, the weight of the submissile and the length of travel. Since the gun length is necessarily short, a propellant must be used which will build up peak pressure very rapidly and fall off rather slowly.

It is very likely that the submissile will have to be restrained until peak pressure is built up so that the full length of travel along the ejection tube can be used to best advantage.

The equation of motion of a submissile during acceleration is

$$m_p \ddot{x} = A_B P_v(x) \qquad (4\text{-}6.1)$$

where, in consistent units,

m_p = projectile mass = W_p/g
$\ddot{x}$ = projectile acceleration
A_B = bore area
$P_v(x)$ = pressure producing velocity (this pressure being a function of the distance traveled)

For purposes of analysis it is assumed that the submissiles are restrained until a peak pressure, P_o, is obtained and that upon release of the submissiles the pressure drops linearly to zero at $x = l$, where l is the total travel. This simplifying assumption is quite obviously not strictly true due to the fact that there is normally a positive pressure present at $x = l$, and the resulting equation will indicate lower submissile ejection velocities than will actually be obtained. However, the developed equation is indicative of what can be expected of the gun tube design. It is further assumed that a 12 per cent pressure loss will be encountered. Under the above assumptions

$$P_v(x) = \frac{P_o}{1.12}\left(1 - \frac{x}{l}\right) \quad (4-6.2)$$
(Reference 4-6.g)

Substituting in equation 4-6.1

$$\ddot{x} = \frac{P_o g}{1.12\, W_p/A_B}\left(1 - \frac{x}{l}\right) \quad (4-6.3)$$

Equation 4-6.3 can be integrated to give

$$\dot{x}^2 = \frac{2 P_o g}{1.12\, W_p/A_B}\left(x - \frac{x^2}{2l}\right) + C \quad (4-6.4)$$

If the following boundary conditions are applied,

at $x = 0$ $\dot{x} = 0$
at $x = l$ $\dot{x} = V_R$,

the following expression for the radial velocity at the end of travel is obtained:

$$V_R^2 = \frac{P_o g l}{1.12\, W_p/A_B}, \quad (4-6.5)$$

For convenience, values from equation 4-6.5 are plotted in Figure 4-66 with W_p/A_B as one parameter.

Since the structural analysis of the gun tube is more or less unfamiliar to the average warhead designer, a standard method of analysis follows. The gun tube or barrel must withstand an internal pressure of P_o. The maximum stresses at the inner surface are therefore

$$\sigma_\theta = P_o\left(\frac{r_o^2 + r_i^2}{r_o^2 - r_i^2}\right) = \text{Tangential Tension} \quad (4-6.6)$$

$$\sigma_r = P_o = \text{Radial Compression} \quad (4-6.7)$$

$$\tau = P_o\left(\frac{r_o^2}{r_o^2 - r_i^2}\right) = \text{Shear} \quad (4-6.8)$$

where:

r_o = outside radius and r_i = inside radius.

Letting $K = \frac{r_i}{r_o}$,

$$\sigma_\theta = P_o\,\frac{1 + K^2}{1 - K^2} \quad (4-6.9)$$

$$\tau = P_o\,\frac{1}{1 - K^2} \quad (4-6.10)$$

Assuming that the tube is made of 4130 steel heat-treated to 200,000 psi (UTS), and using a safety factor of 1.5 based on yield stress, equations 4-6.7, 4-6.9, and 4-6.10 become

$$\sigma_r = \frac{165,000}{1.5} = P_o, \quad (4-6.11)$$

$$\sigma_\theta = \frac{165,000}{1.5} = P_o\,\frac{1 + K^2}{1 - K^2}, \quad (4-6.12)$$

and

$$\tau = \frac{115,000}{1.5} = P_o\,\frac{1}{1 - K^2} \quad (4-6.13)$$

where:

165,000 psi = tensile yield stress of the material

115,000 psi = ultimate shear strength of the material

For convenience, values from equations 4-6.12 and 4-6.13 are shown graphically in Figure 4-67.

Since only one ejection need be considered, and that of short duration, stresses up to the yield stress of the gun tube material may be used. Similar curves for 4130 heat-treated to 180,000 psi are shown in Figure 4-68, and for aluminum 61ST in Figure 4-69. These latter curves do not include the 1 safety factor.

Blast Method In blast type ejection the submissiles are ejected directly by the explosion of the propellant. The major difference in this method as compared to the gun tube method is the use of a central blast chamber requiring only one safety and arming mechanism, with conservation of weight by the elimination of backup structure. With ejection methods utilizing this central blast chamber, accelerating forces are applied to the submissiles over a short distance of travel; therefore, in order to achieve reasonable ejection velocities, large accelerating forces are necessary.

There is little analytical information available in this area. The short period of time that blast ejection warheads have been studied and the very large number of variables involved in the design have thus far made it impossible to develop formulas or graphs from which to design the ejection configuration in a relatively simple manner.

The simplest blast ejection method is to eject the submissiles as if they were, in effect, large preformed fragments. This method was developed by the Rheem Manufacturing Company for use in the T-46 Cluster warhead and the results are reported in Reference 4-6.h. The submissiles consisted of spherical balls weighing approximately 4.4 pounds each, arranged in rings around the warhead. They are ejected by the explosive charge contained in an aluminum tube in the center of the warhead as shown in Figure 4-70. A plot of charge-to-metal ratio *(c/m)* versus initial velocity of the submissile is shown in Figure 4-71. This can be used to obtain a rough approximation of the charge required for a given velocity of the submissile. The velocity control has been found to be quite sensitive to the standoff distance between the charge and submissile. The *c/m* ratio in this design is based on an individual ring of submissiles and the explosive core directly inside the individual ring.

Figure 4-72 shows the submissiles arranged in a prototype T-46 warhead. Figure 4-73 indicates the typical flight pattern of these submissiles directed at the target for a static firing of the warhead. It can be seen that the velocity of the individual submissiles varies between 200 and 400 feet per second.

Although the piston-cylinder type has been proven adequately, it is not favored by many in this field as its high weight is undesirable and the close manufacturing tolerances needed are highly unfavorable to large scale production.

The most promising method developed to date, but as yet unproven in an actual warhead, is the modified ported chamber with a convoluted expanding liner.

Design of Submissiles

Unstabilized Submissiles One illustrative type of unstabilized submissile consists of the trapezoidal case, which carries two internally threaded rings, the ejection tube, the fuze, and the filler cap or plug, the last two items being screwed into the two rings.

The thickness of the case is primarily determined by the structural requirements to withstand the ejection forces. It has been found that, for an overall weight of approximately 5 to 6 pounds and initial ejection velocities up to 350 fps, the case can be made of 5052-H34 aluminum alloy sheet of .064 thickness with the sheet drawn to the wedge-shape and the outer cylindrical face welded or fused into place. The mounting rings for fuze and filler cap are usually made of aluminum and are welded into this face. The filler cap and fuze cap also can be made of aluminum.

The ejection tube is usually made of steel, has a closed outer end and is provided on the other end with a flange external to, and attached to the inner face of the submissile. This flange serves to transmit the force of the explosion to the inner face of the submissile. An "all-ways fuze" (for unstabilized submissiles) is provided for arming, detonation, and self-destruction.

The filling of the explosive charge is accomplished through a filler cap screwed into its mounting ring. The ejection tube, fuze and filler cap and rings project into, and subtract from, the inside space of the case with the remainder of the case filled with the charge. This charge is generally of the HBX type.

Another illustrative type of unstabilized submissile consists of a spherical case, shown in Figure 4-74, which is fastened directly to the warhead structure by a Dzus fastener.

The thickness of the case is determined both by the structural requirements to withstand the ejection forces, and the necessity for penetrating the target structure. The submissile weighs approximately 4.42 pounds and is 4.24 inches in diameter. Ejection velocities range up to 400 fps. The entire submissile is made of steel, and it utilizes an "all-ways fuze". The charge used is of the HBX type.

Stabilized Submissiles The optimum design stabilized submissile is a directionally stable body capable of rapidly damping the ejection angular transients to a relatively small magnitude, and one whose shape is compatible with packaging restrictions imposed by the warhead compartment. The submissile and stabilization device must also be strong enough to withstand the ejection forces. A number of aerodynamic arrangements have been considered and sketches of them with the investigators' remarks are shown in Figure 4-75 (Reference 4-6.c).

As indicated in Figure 4-75, the fixed fin plus viscous damped elevon combination was, for the same weight and size of surfaces, far superior to the other devices. However, other investigations have shown theoretically that, with a well designed drag tube or drag plate, adequate aerodynamic stabilization is possible. No experimental evidence supporting or disproving this was found.

The shape of the submissile body is governed by its penetration characteristics and its adaptability to efficient packaging in the warhead space available. However, since the clusters are generally limited by weight rather than volume, the problem can be resolved into one of determining the best penetration characteristics. On the basis of firing test data on warheads under development, the T-214 rocket warhead appears to have superior characteristics. As shown in Table 4-8, the test data indicates that the T-214 will penetrate typical aircraft structure at high obliquity angles even when yawed as much as 30 degrees (Reference 4-6.c). The range of striking velocities over which the high obliquity penetration tests were conducted was between approximately 700 and 2700 feet per second. Other warhead types are also shown for comparison.

Typical calculations on the use of drag tube and drag plate stabilized submissiles are shown in detail in Reference 4-6.a. These two methods are recommended on the basis of being likely to perform with minimum developmental effort. The referenced analysis indicates that both the drag plate and the drag tube are capable of aerodynamically stabilizing the submissile. The question of whether the stability thus produced is adequate depends on the following factors: (1) the maximum yaw at collision with the target capable of being tolerated by the submissile fuze, (2) the yawing influences present at ejection, (3) the fact that the drag plate or drag tube is in the turbulent wake which was neglected in the analysis and (4) the submissile slow-down in velocity caused by the drag configuration. Experimental verification is needed before any definite predictions can be made on this type of stabilization.

Since fixed fins which would be ideal for the submissile stabilization are difficult to package, the use of folding fins has been considered. An extensive development program was conducted by the Armour Research Foundation Reference 4-6.d on the use of folding fin type submissiles, called Sprites, for the Sparrow I warhead. The referenced report represents the most complete investigation to date, and contains a thorough analysis of the entire development program.

The final design evolved in the Armour program is shown in Figure 4-76 and has been proven highly satisfactory on sled tests. Tests under actual operating conditions have not yet been performed. The design makes use of a folding fin with a single axis of rotation, and having the following properties: (1) when folded, the fin lies flush with the submissile surface; when unfolded the fin forms part of a conventional configuration, i.e., one in which the surfaces of the various fins intersect on a common line; and (2) the axis of rotation is such that ejection setback forces are sufficient to open the fins with extreme rapidity. Therefore, no springs or other devices are needed to actuate the fin. A detent fin lock, shown in Figure 4-77, is used to lock the fins in the open position after submissile ejection. The fin rotational velocity prior to locking can be controlled by the use of a soft aluminum washer under the nut on the pivot pin in conjunction with locking the nut to the fin post as shown in Figure 4-77. Rotation of the pin thereby causes the nut to screw further onto the pivot pin, and extrudes the aluminum washer. The energy absorbed by the washer results in a slower fin rotational velocity and provides more positive locking. The amount of energy absorbed can be controlled by varying the initial nut torque.

Welding has been found unsatisfactory for the manufacture of the sprites, but an investment casting process using frozen mercury patterns and AISI 410 steel has given good results. Tool steel, heat-treated to 300,000 psi has been found satisfactory for the detents.

No conclusive work of a comparative nature is available in the stabilized submissile field. An attempt has been made to show the work done to date and to provide the designer with a reasonable basis on which to proceed with his particular warhead. At the present state-of-the-art of stabilized submissile design an extensive test program must be conducted to prove the reliability of any type of stabilization system selected.

Design of Support Structure Regardless of the method of ejection used, a central structure for carrying the submissiles must be incorporated. This structure must be capable of withstanding the radial reaction forces of the submissile ejections and also the fore and aft inertia forces acting on the submissiles during the launching of the missile.

The most economical configuration for this structure is a tubular form, strengthened locally at each row of submissiles. This tube should contain fittings at both ends for attaching the warhead to the actual structure of the missile, preferably in such a way that the entire warhead assembly can be readily assembled to or disassembled from the missile, and transported as an assembly.

Since this structure is so important to the success of the missile it should be carefully analyzed for its structural integrity. For any tubular structure the following equations may be used:

Maximum Bending Moment Between Loads:

$$M_B = \tfrac{1}{2} F R_G \left(\frac{1}{\sin \theta_f} - \frac{1}{\theta_f} \right) \qquad (4\text{-}6.14)$$

Maximum Tangential Compression Between Loads:

$$T_B = \tfrac{1}{2} F \left(\frac{1}{\sin \theta_f} \right) \qquad (4\text{-}6.15)$$

Maximum Bending Moment at Loads:

$$M_A = \tfrac{1}{2} F R_G \left(\cot \theta_f - \frac{1}{\theta_f} \right) \qquad (4\text{-}6.16)$$

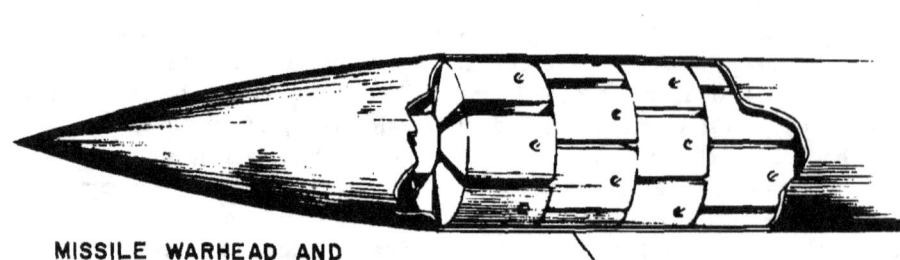

MISSILE WARHEAD AND
SUBMISSILE ARRANGEMENT

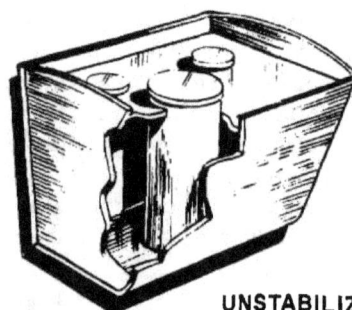

UNSTABILIZED SUBMISSILE

SUBMISSILE AND EJECTOR ARRANGEMENT

Figure 4-51. Unstabilized Submissiles - Typical Arrangement
(Gun Tube Method)

Total Tangential Compression at Loads:

$$T_A = \tfrac{1}{2} F \cos \theta_f \qquad (4\text{-}6.17)$$

where: F = force in pounds,

R_G = radius of centroid of tubular cross section and

θ_f = 1/2 angle between forces.

See Figures 4-78 and 4-79.

If the design does not readily lend itself to conventional structural analysis, one reliable method of determining the actual stresses which will be developed is by the use of "Stresscoat", a nondestructive brittle coating which cracks perpendicular to the maximum principal stress in the coated surface. This method has been successfully employed in the past, and is completely described in Reference 4-6.d.

Design of Retention System It is necessary to retain the submissiles in place, both in actual flight of the parent missile, and in handling and transport of the warhead assembly. Various methods have been investigated and used. These consist primarily of shear pins and retaining bands. The shear pin method involves the use of shear pins in a conventional manner, and requires no explanation here.

A simple and satisfactory method of retention is by means of a steel strap arranged tightly around the outer faces of the submissiles of each row. This strap may be of a commercial type used for banding crates. Tests conducted for the Nike I missile by Aircraft Armaments, Inc., found a strap size of 3/8 x .010 was sufficient to withstand an ultimate load factor of 50 g, with a submissile weight of 4.6 pounds. The removal of this strap is accomplished by the ejection forces on the submissiles. About 1 to 2 per cent of the peak firing pressure was utilized to break the strap. It is therefore apparent that no means other than missile ejection need be supplied for strap removal. This method is by far the simplest and least expensive to fabricate, and is highly recommended. Figure 4-80 illustrates a typical installation utilizing this method, which can be used for retaining both stabilized and unstabilized submissiles.

Design of Obstruction Removal Devices In order that the submissiles may have an uninterrupted path once they have been ejected, certain obstacles have to be removed prior to ejection. These can include the skin of the parent missile around the warhead compartment, any longitudinal stringers, longitudinal ribs or fins external to the missile skin, electric wiring carried fore and aft inside the skin or in external ribs, and piping carried in a similar fashion.

In some cases the missile skin may be thin enough to be blown away by the force of the ejected submissiles without a substantial decrease in velocity, but generally it is advisable to have a positive method of skin removal, particularly with the thicknesses used when the skin forms a structural member. This removal can be accomplished most effectively by arranging a harness of linear shaped charges as shown in Figure 4-81, which act to cut the skin circumferentially at the fore and aft ends of the warhead compartment, and longitudinally into several sections. The explosive is contained in brass channel sections and retained therein by adhesive tape. Detonators and boosters are arranged at each junction of the channels, and are initiated simultaneously with the ejection firing of the submissiles.

One of the disadvantages of the use of linear shaped charges is that the internal missile components are subject to damage from the back blast unless adequate shielding is employed, or proper techniques used in designing the explosive charge. The blast is essentially unidirectional, but the back blast plus side spray and metallic fragments are sources of damage to improperly protected equipment. This back blast effect can be reduced by retaining the charge, on the side away from the surface to be cut, with foam plastics, foam rubber or solid rubber.

Detonating cord can also be used for cutting the missile skin, but its pressure pattern appears to cause damage over a wider area than the linear shaped charge. However, stringers and external ribs or fins can be cut and blown away more easily by firing lengths of detonating cord attached adjacent to the objects to be cut. The designer must make provisions to shield and protect any equipment in the area of these charges.

Ribs or fins that are bolted externally to the surface of the missile can be removed by using explosive bolts for attachment. These bolts are hollowed out and a standard detonator inserted. When fired, the bolt fails and the parts will separate. Control of the point at which the bolt fails may be obtained by undercutting. The blast damage may be reduced by the use of a sleeve around the bolt.

For severing piping and electric wiring detonating cord may be used, but if the piping or wiring is large, the use of a guillotine for cutting has been proved satisfactory. The guillotine consists of a hardened knife operated by an explosive charge that is initiated at the time of warhead detonation. The guillotines are used at only one end of the cabling or piping. The blast from firing this type of device is small, and with proper orientation will cause no damage to other components. Figure 4-82 is an illustration of the top of a missile with the fairing removed showing the installation of the guillotine. Figure 4-83 illustrates the type of cutting action obtained with the guillotine.

If the charge is placed against an outside skin of the missile, the designer must consider the possible skin temperature during flight. Some types of explosives will begin to fume at low temperatures (e.g. RDX detonating cord at 325°F), and since the speed of the missile is sometimes great enough to raise the skin temperature a substantial amount, consideration must be given to preventing the charge from pre-igniting. This can be accomplished either by use of an insulator, by making the charge stand off from the surface a short distance, or by selecting an explosive not affected by the temperatures encountered.

Summary of Fuzing Requirements Once the design is final, a summary data sheet should be prepared for the benefit of the fuze designer to permit him to effect a fuze design which will be compatible with the warhead. The following data is required:

    A. Warhead Fuze
        (1) Initial Ejection Velocity of Submissiles
        (2) Number and Pattern of Submissiles
        (3) Type of Ejection System
        (4) Type of Ejection Charge (Blast)
        (5) Detail Design Drawings (Warhead)
    B. Submissile Fuze
        (1) Type of Submissiles
        (2) Weight of Charge
        (3) Type of Charge
        (4) Detail Design Drawings (Submissile)
        (5) Type of Fuze Action Required

Summary of Design Data At the conclusion of the design procedure one should prepare a summary of all the pertinent data evolved. This should include the following items:

    (1) Total Weight
    (2) Detail Design and Installation Drawings
    (3) Explosive
        (a) Material
        (b) Weight
    (4) Charge-to-Metal Ratio (c/m)
    (5) Submissiles
        (a) Number
        (b) Total Weight
        (c) Individual Weight
        (d) Design Size and Shape
        (e) Initial Velocity
        (f) Pattern
        (g) Casing Material and Thickness
    (6) Ejection System
    (7) Backup Structure
    (8) Mounting System in Missile
    (9) Intended Operation of Weapons System
    (10) Location of Center of Gravity

4-6.3. References

4-6.a "Preliminary Design of a Cluster Warhead for the Bomarc Guided Missile", Armour Research Foundation, ASTIA AD-46 907, November 1954.

4-6.b "Development of a Cluster Type Warhead for the Bomarc Missile", Aircraft Armaments, Inc., Report No. ER-337, August 1955.

4-6.c "Feasibility Study of a Stabilized Submissile Cluster Warhead for Nike I", Bell Telephone Laboratories, ASTIA AD-81 337, May 1955.

4-6.d "Development of a Guided Missile Warhead", Armour Research Foundation, ASTIA AD-50 251, October 1954.

4-6.e "Development of a Cluster Type Warhead for the Nike B Missile", Aircraft Armaments, Inc., Report No. ER-442, July 1954.

4-6.f "Cluster Warhead for Nike Missile", Aircraft Armaments, Inc., Report No. ER-301, August 1953.

4-6.g "Effectiveness of a Cluster Type Warhead as an Antiaircraft Weapon", Aircraft Armaments, Inc., Report No. ER-173, October 1952.

4-6.h "T-46 Cluster Warhead", Rheem Manufacturing Company, ASTIA AD-88 823, December 1955.

4-6.4. Bibliography

(1) "Final Report, Cluster Warhead for Nike I Missile", Aircraft Armaments, Inc., Report No. ER-388, August 1955.

(2) "Feasibility Program Cluster Type Warhead for the Bomarc Missile", Aircraft Armaments, Inc., Report No. ER-676, October 1955

(3) "Final Report, Cluster Warhead for Nike B Missile", Aircraft Armaments, Inc., Report No. ER-865, June 1956.

(4) "The Selection of the Optimum Charge to Metal Weight Ratio for Warheads for Antiaircraft Missiles", U. S. Naval Ordnance Laboratory, ASTIA AD-24 429, October 1953.

(5) "Feasibility Study of Bomarc Submissile Fuze", Remington Rand, ASTIA AD-27 828, January 1954.

(6) "T-46 Cluster Warhead Progress Report", Rheem Manufacturing Company, ASTIA AD-75 209, August 1955.

(7) "T-46 Cluster Warhead", Rheem Manufacturing Company, ASTIA AD-91 463, January 1956.

(8) "T-46 Cluster Warhead", Rheem Manufacturing Company, ASTIA AD-85 169, November 1955.

(9) "T-46 Cluster Warhead", Rheem Manufacturing Company, ASTIA AD-88 823, December 1955.

(10) "Single Shot Probabilities for Bomarc Cluster Warheads", R. L. Simmons, BRL Memo, Report No. 800, ASTIA AD-42 517, March 1954.

(11) "Single Shot Probabilities for Nike Cluster Warheads", R. L. Simmons, BRL Memo, Report No. 797, ASTIA AD-43 060, May 1954.

(12) "Design and Development of Fuze Type T-1412 for Cluster Warheads", Remington Rand, ASTIA AD-82 201, August 1955.

(13) "Survey of Guided Missile Warheads", Haller, Raymond and Brown, Inc., Report No. 91-R-5, September 1956.

4-7. SHAPED CHARGE WARHEADS

4-7.1. Detail Design Steps

Step
No.   Detail Design Steps
1. Establish the Penetration Required to Enter the Target.
2. Determine the Allowable Weight and Envelope.
3. Establish the Type of Confinement (determined by the structure of the delivery vehicle).
4. Choose the Explosive with the Highest Detonation Pressure (Comp. B as of 1958).
5. Choose the Length/Diameter Ratio.
6. Design the Liner.
7. Prepare Summary of Fuze Data.
8. Prepare Summary of Design Data.

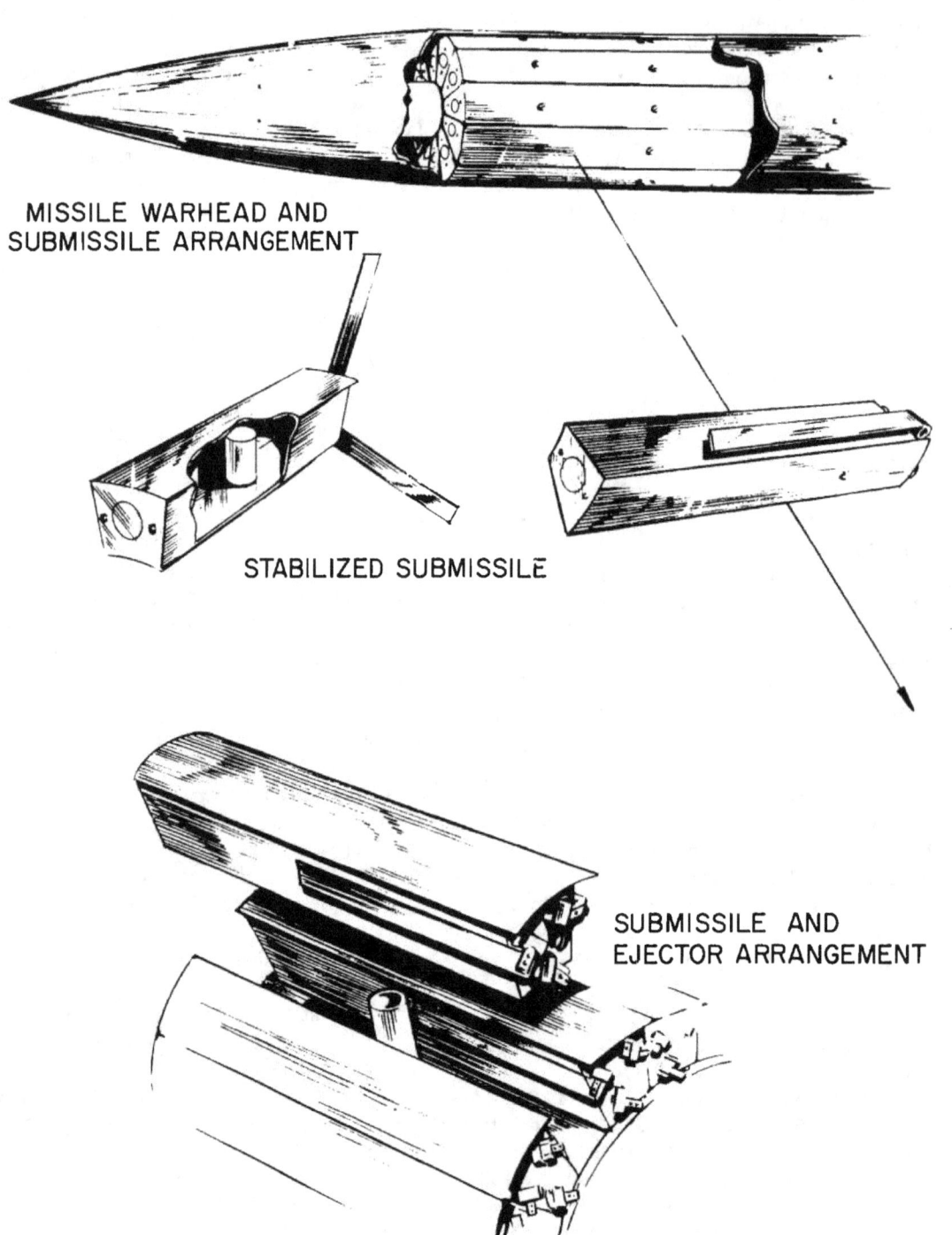

Figure 4-52. Stabilized Submissiles - Typical Arrangement

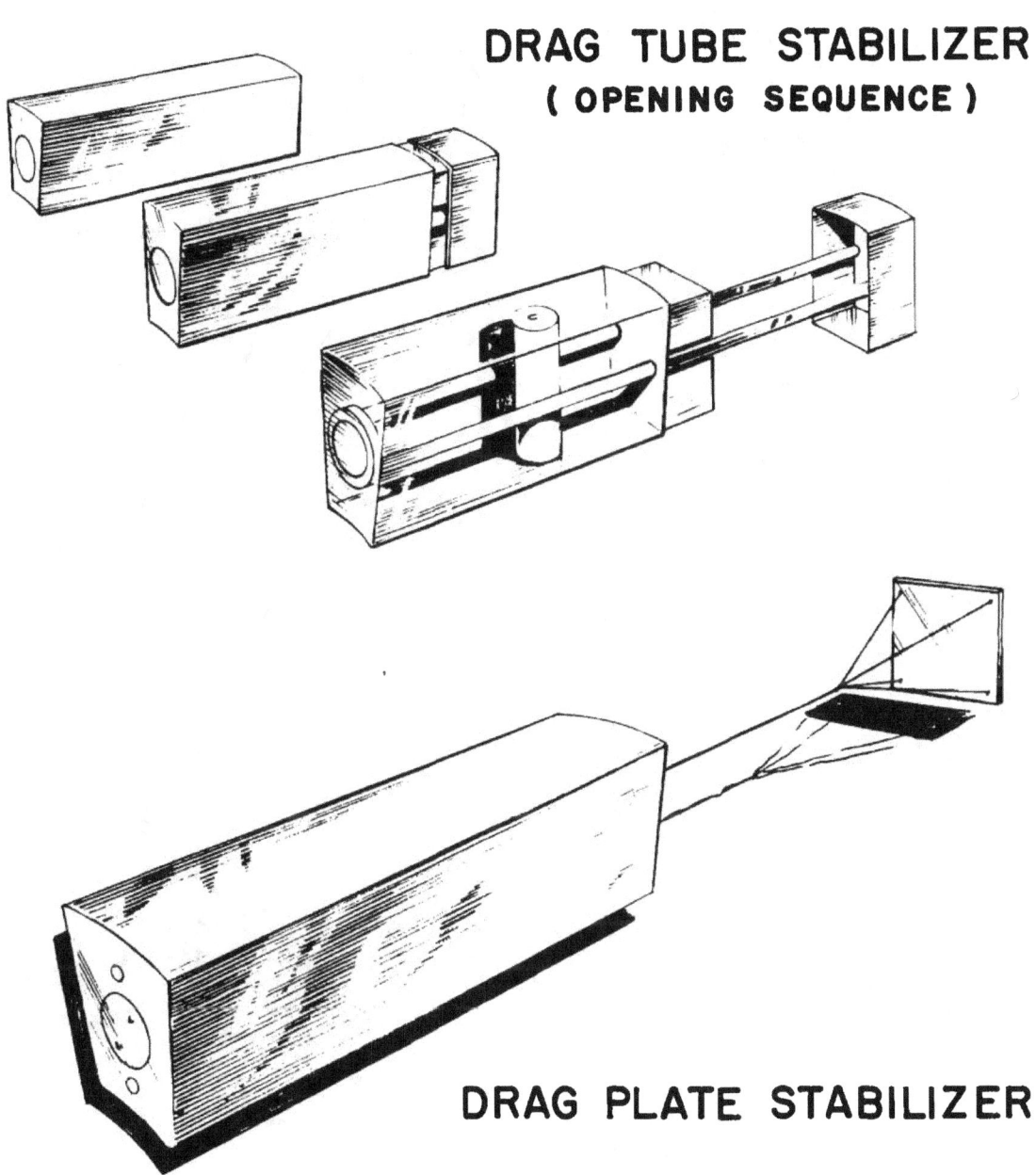

Figure 4-53. Drag Tube and Drag Plate Stabilizer

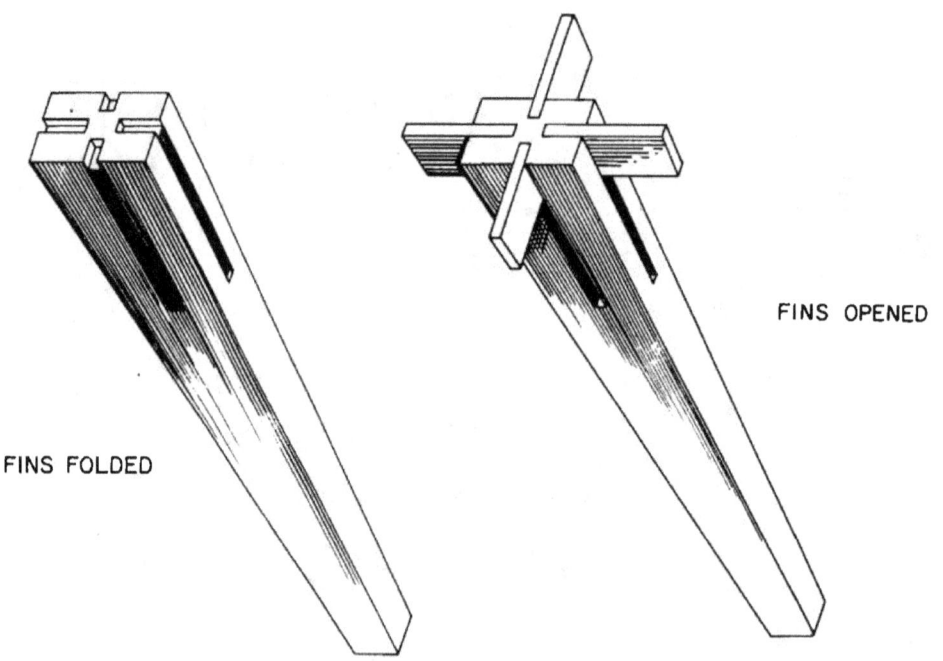

Figure 4-54. Folding Fin Stabilizer

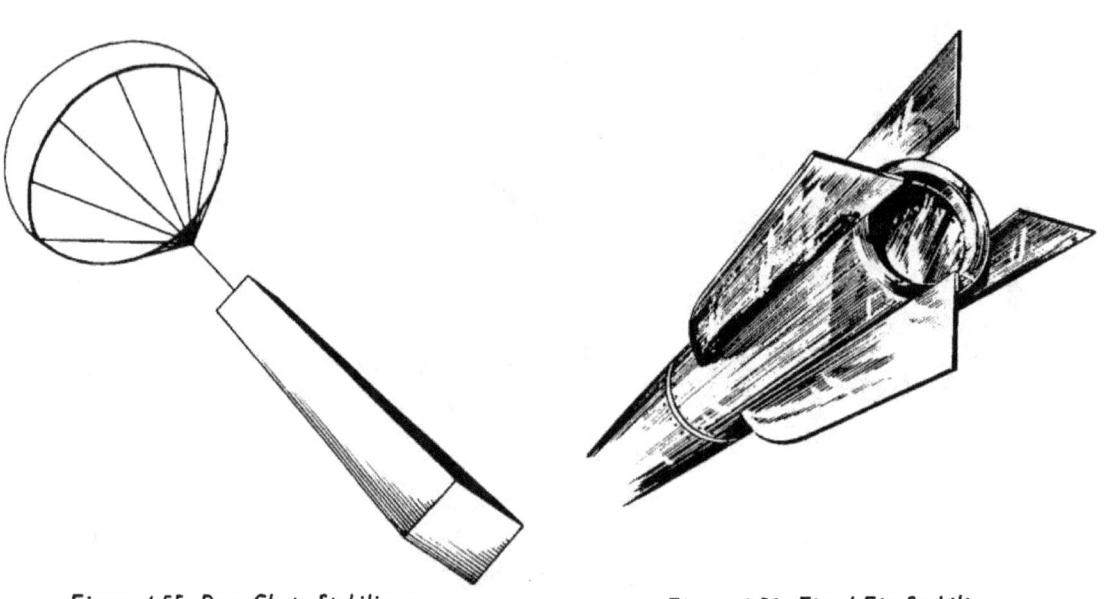

Figure 4-55. Drag Chute Stabilizer

Figure 4-56. Fixed Fin Stabilizer

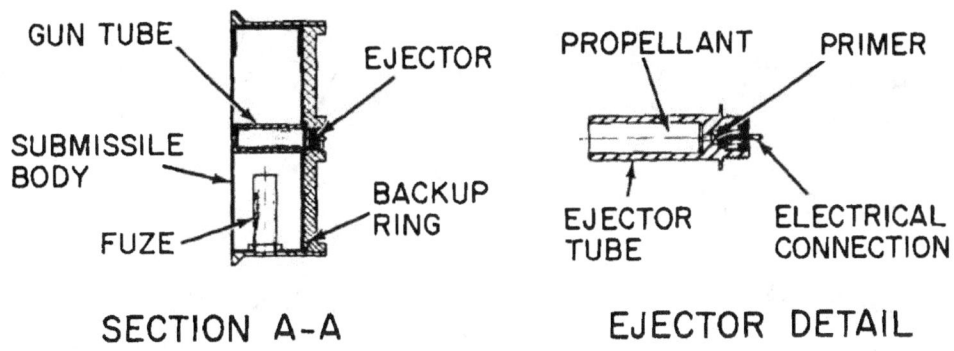
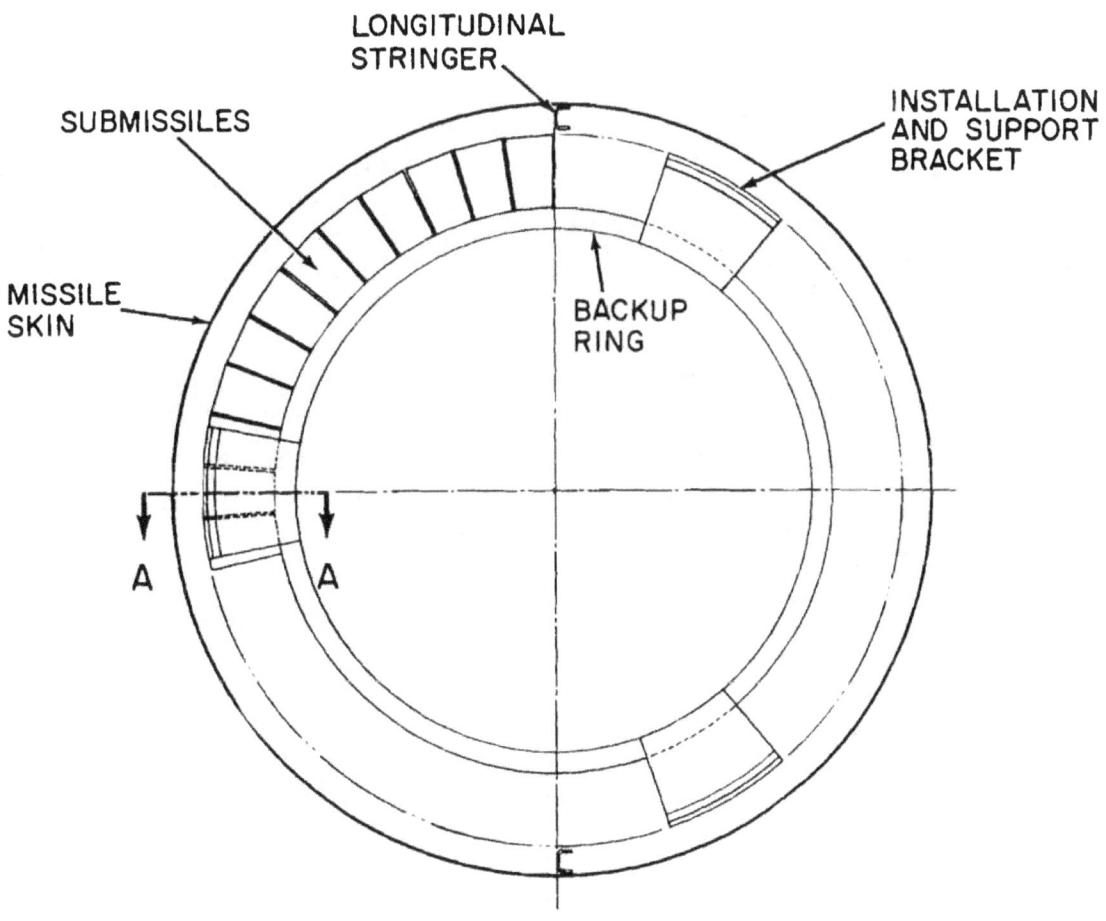

Figure 4-57. Integral Ignition System (Gun Tube Method)

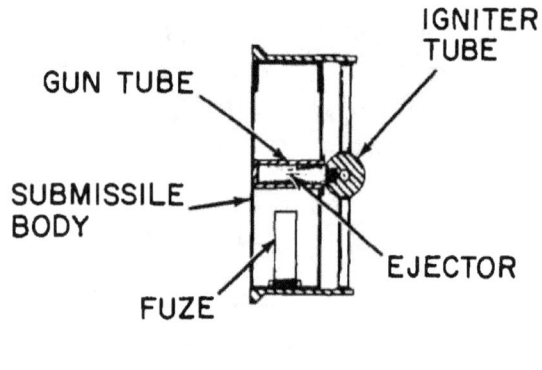
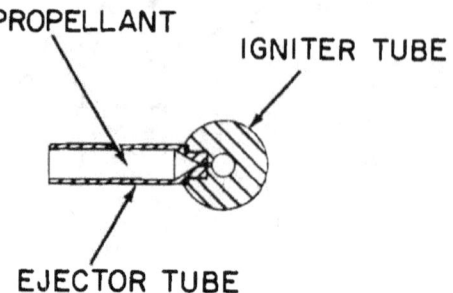

SECTION A-A   EJECTOR DETAIL

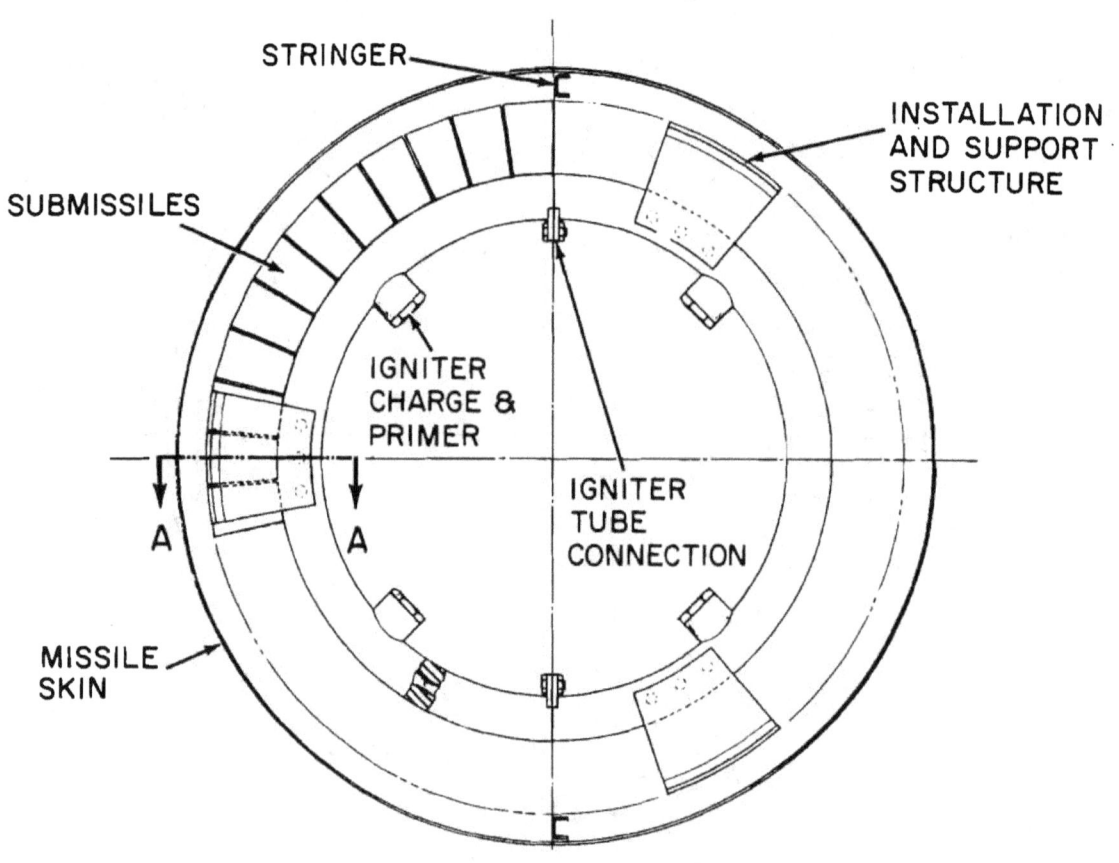

Figure 4-58. Central Ignition System
(Gun Tube Method)

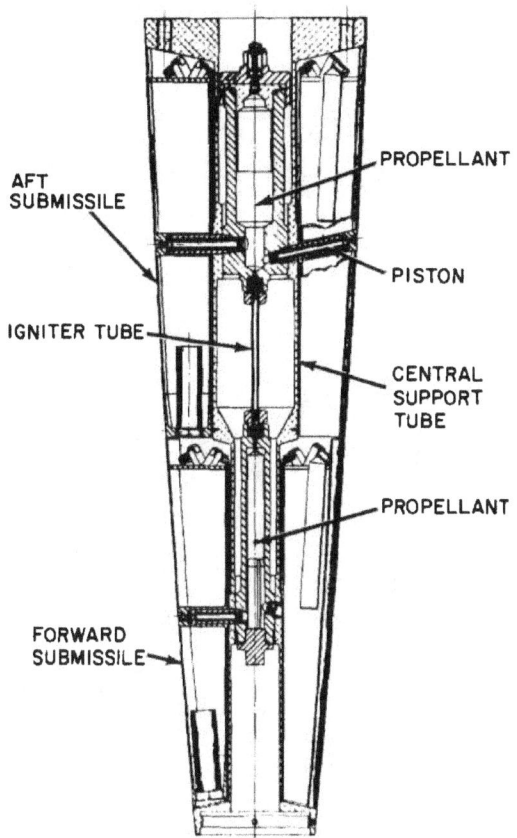

Figure 4-59. Piston Type Ejection

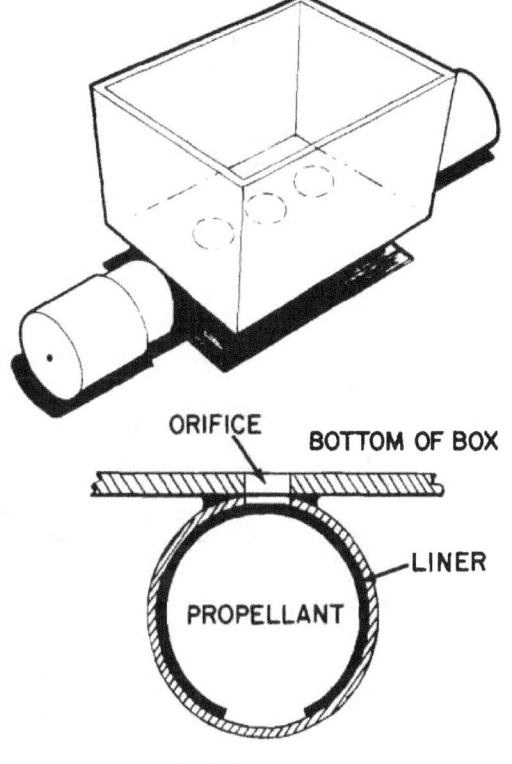

CROSS SECTION OF STEEL CASE

Figure 4-60. Blast Ejection - Intermediate Liner

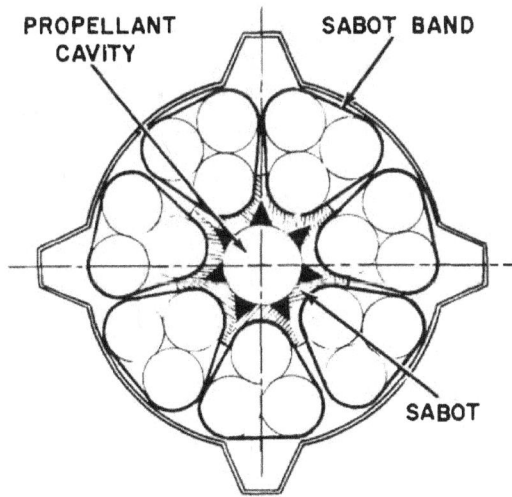

21 SUBMISSILES
SHOWN IN CLUSTERS OF THREE

Figure 4-61. Blast Ejection - Segmented Chamber

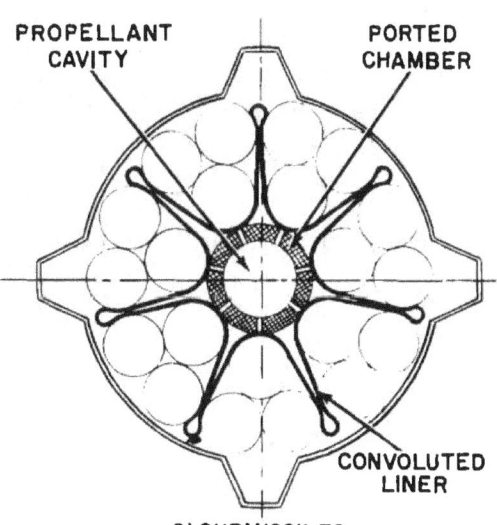

21 SUBMISSILES
SHOWN IN GROUPS OF THREE

Figure 4-62. Blast Ejection - Convoluted Liner

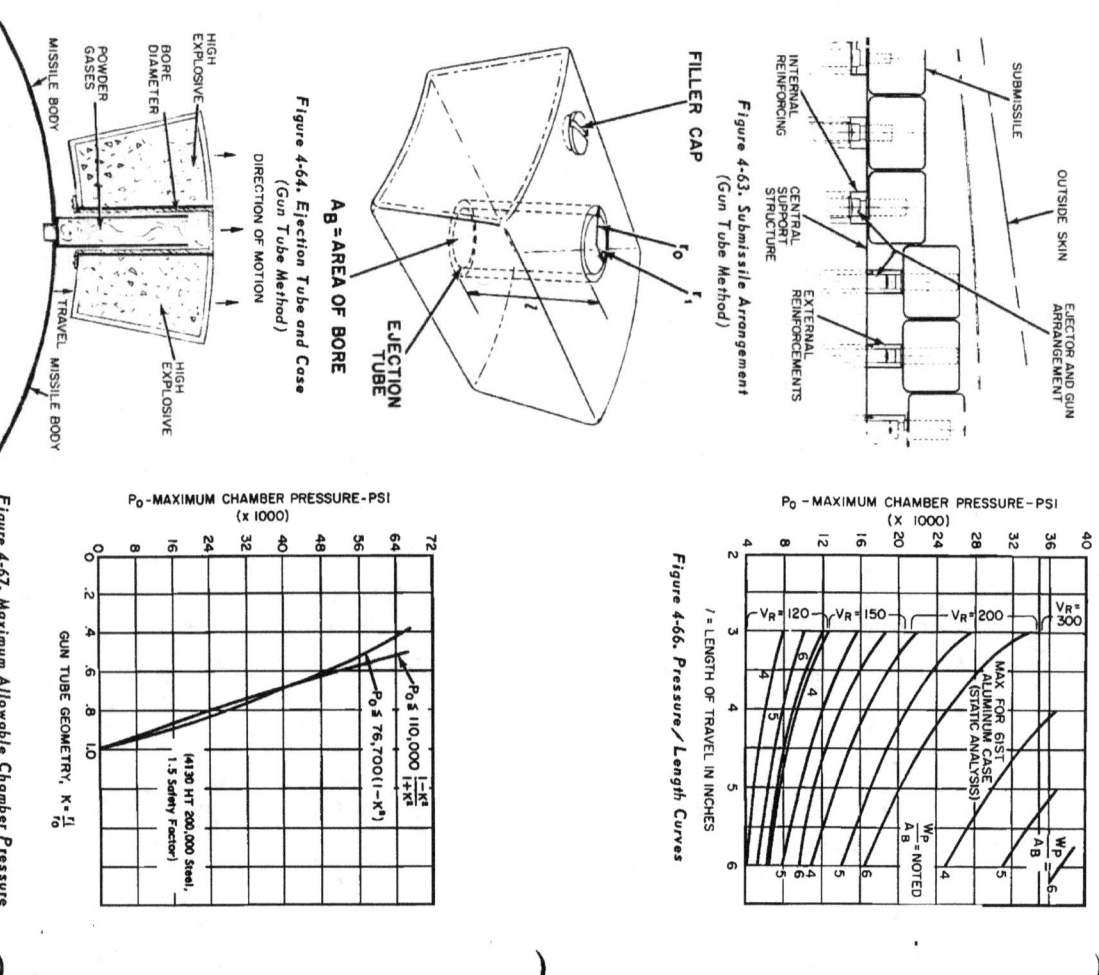

Figure 4-63. Submissile Arrangement (Gun Tube Method)

Figure 4-64. Ejection Tube and Case (Gun Tube Method)

Figure 4-65. Gun Tube Ejection

Figure 4-66. Pressure/Length Curves

Figure 4-67. Maximum Allowable Chamber Pressure Vs. Gun Tube Geometry

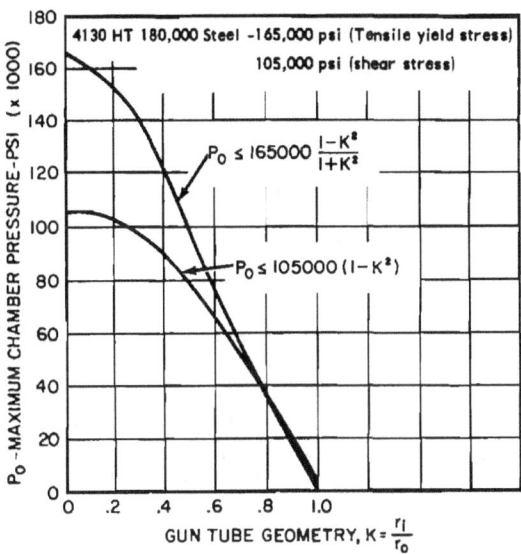

Figure 4-68. Maximum Allowable Chamber Pressure Vs. Gun Tube Geometry

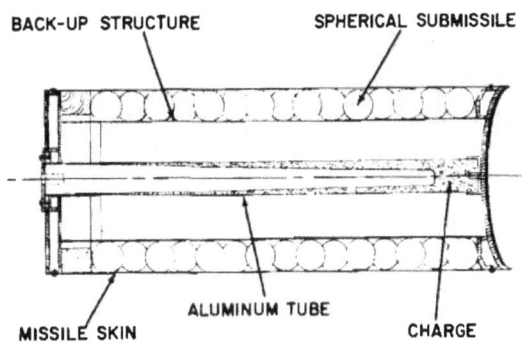

Figure 4-70. Spherical Submissile Warhead

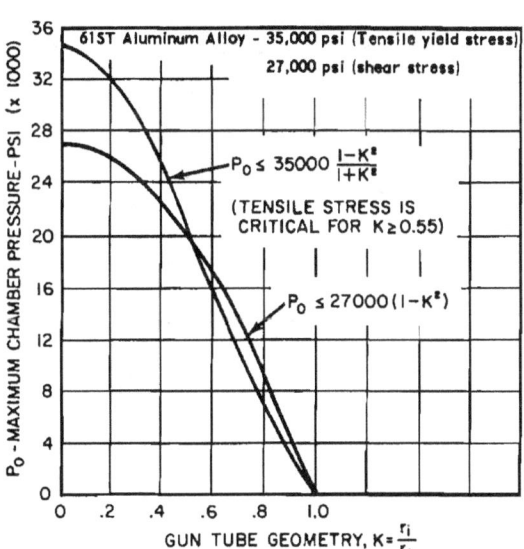

Figure 4-69. Maximum Allowable Chamber Pressure Vs. Gun Tube Geometry

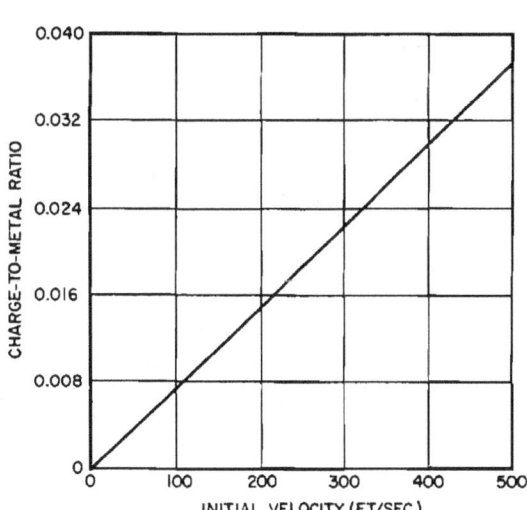

Figure 4-71. Charge-to-Metal Ratio Vs. Velocity for T-46 Cluster Warhead

119

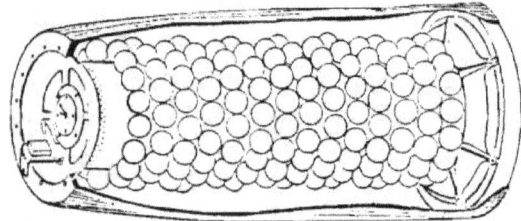

Figure 4-72. T-46 Prototype Warhead

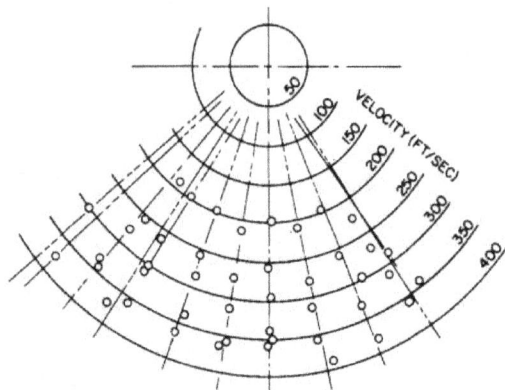

Figure 4-73. Typical Submissile Flight Pattern

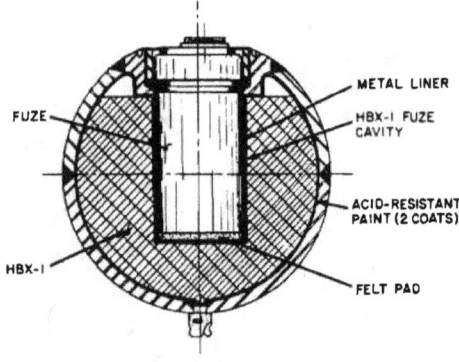

Figure 4-74. Spherical (Rheem) Submissile

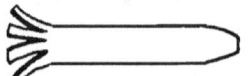

Fixed Base: Poor inherent Damping.

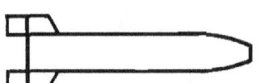

Fixed Fins with Damped Elevon: Most promising.

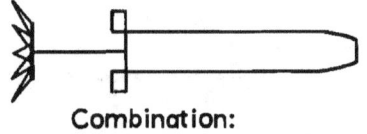

Combination: Needs further study.

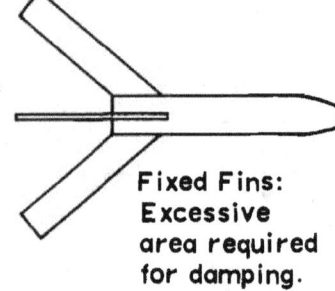

Fixed Fins: Excessive area required for damping.

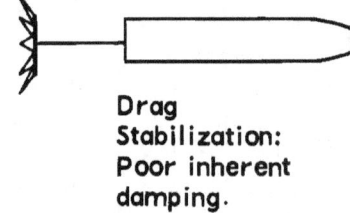

Drag Stabilization: Poor inherent damping.

Figure 4-75. Types of Stabilization

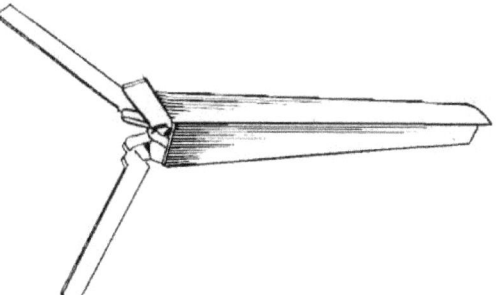

Figure 4-76. Fin Stabilized Submissile

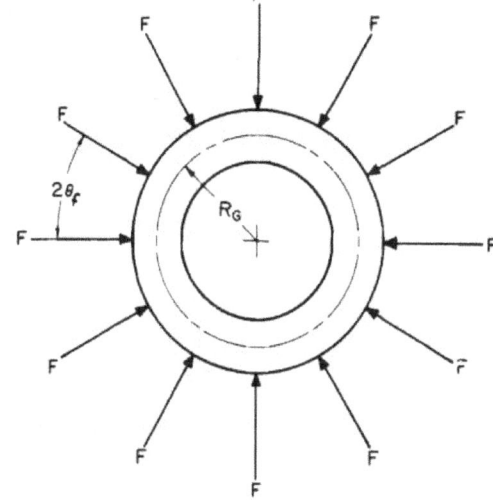

Figure 4-79. Tube Analysis

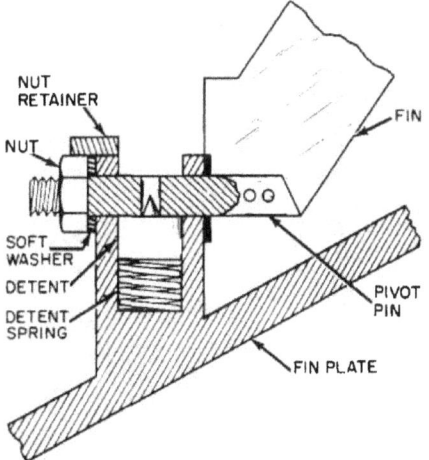

Figure 4-77. Fin Lock

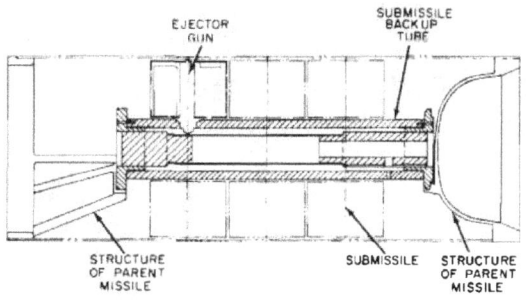

Figure 4-78. Support Structure

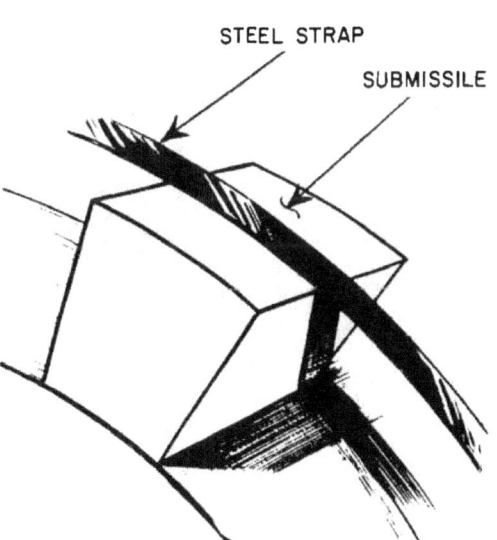

Figure 4-80. Submissile Retention

121

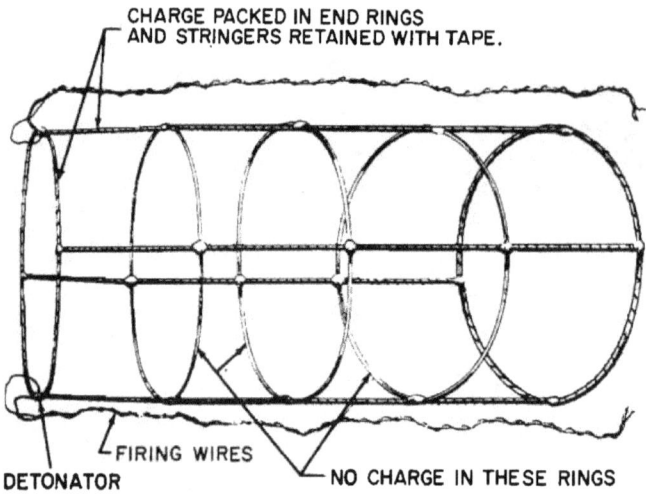

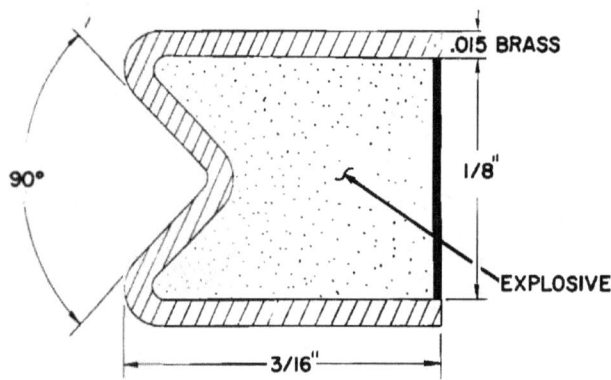

Figure 4-81. Skin Removal Harness

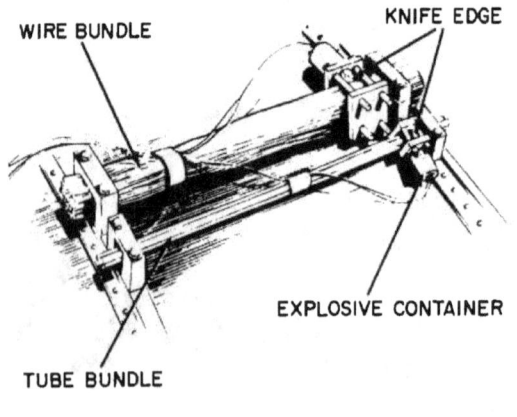

Figure 4-82. Guillotine Installation

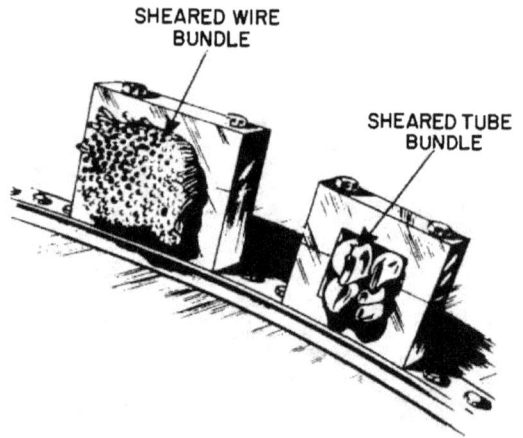

Figure 4-83. Guillotine Effect

The exact order of the design procedure may vary depending upon the viewpoint of the designer and, even more, on the military requirements which often fix certain parameters in advance.

4-7.2. Detail Design Data

Liner Design The design of the shaped charge warhead liner is most important. (See Figure 4-84 for shaped charge nomenclature.) The variables to be considered in the design of the liner are (1) liner diameter, (2) liner material, (3) liner profile, and (4) thickness of liner material. The performance of the warhead against various targets is primarily dependent upon an intimate interrelationship of these variables among themselves and with standoff.

The upper limit on liner diameter is obviously established by the diameter of the warhead compartment, which is determined by the missile system designer. The performance of a given diameter copper cone measured in terms of its penetration of armor steel, with all other variables being optimum, is given by the empirical relationship:
(Reference 4-7.b)

$$D = \frac{T + 2}{5} \quad (4-7.1)$$

Where D = cone diameter in inches
T = thickness of armor in inches

The above equation does not necessarily describe the overall optimum performance against armor because the optimum standoff condition cannot usually be realized. The penetrating ability in terms of the cone diameter for various steel lined cones against concrete as a function of standoff is presented in Figure 4-85 and Figure 4-86. Shaped charges fired against aircraft at long standoff distances of the order of 100 to 150 feet should have a diameter of at least 6 inches to severely damage the target. (Reference 4-7.e).

The choice of liner material involves one of the basic decisions in shaped charge design. For short standoffs, the order of penetration ability for liner materials is copper, aluminum,

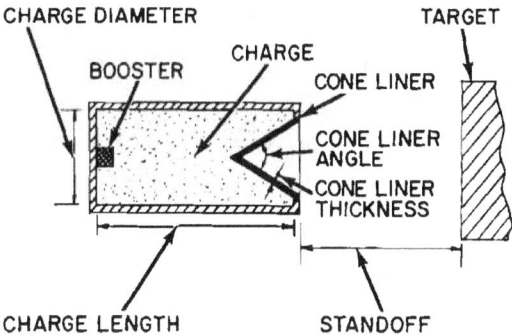

Figure 4-84. Shaped Charge Nomenclature

steel, zinc, lead, and glass. Most designs use copper, aluminum, or steel. Oxygen-free electrolytic copper is considered the best choice when maximum penetration is desired at small standoffs. When choosing between alloys and grades of aluminum and steel, note that the most ductile will yield maximum penetration. Copper liners will give the greatest penetration of steel targets at standoffs of from 1 to 3 charge diameters. See Figure 4-87. The optimum standoff for aluminum liners against steel targets is larger than for copper liners, i.e., about 5 to 7 charge diameters. See Figures 4-88 and 4-89. The optimum standoff for aluminum liners against steel and concrete targets is similar to that for copper liners, that is 1 to 3 charge diameters. See Figures 4-85 and 4-86.

The utilization of the warhead is a determining factor regarding the liner material to use. If maximum penetration is desired at short standoff, the liner material should be copper. However, if the warhead has greater penetration than is required, it would be wise to consider the use of an aluminum or a laminated liner of two metals. By using either of these types of construction, there will be a small loss in penetration, but, behind the target penetrated, the lethal effects will be increased due to special incendiary conditions. If the target is an aircraft, the best material to use for the liner is aluminum, since the greatest amount of damage to low density targets will occur with low density liners. The relative

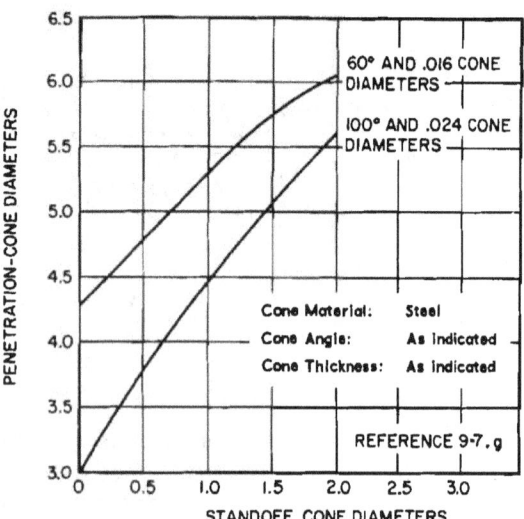

Figure 4-85. Penetration Vs. Standoff; Cone Angle and Cone Thickness Against Concrete

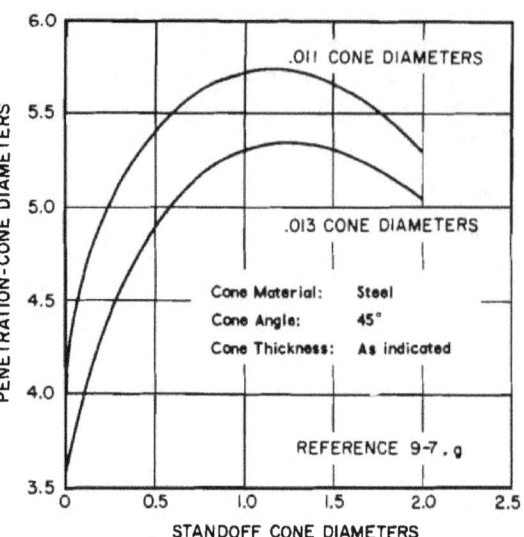

Figure 4-86. Penetration Vs. Standoff and Cone Thickness Against Concrete

penetration capabilities of various liner materials on a unit density basis, taken from reference 4-7.a is presented in Table 4-9. For Table 4-9, the standoff distance was large enough so that the copper jet was beyond maximum penetrability.

Cone apex angles between 40 and 60 degrees give good performance at the standoffs usually associated with surface targets; i.e., 2 to 4 cone diameters. This range of cone angles is used in both spin compensated and nonspin compensated warheads. In spin compensated warheads at low rates of spin, experimental data indicate that a cone whose opening angle changes provides better penetration at shorter standoffs than single-angle cones.

Increased penetration of surface targets can be achieved by well made cones utilizing smaller cone angles of 20 to 30 degrees at standoffs below about two cone diameters. At best, however, the improvement in performance achieved from the smaller cone angles is only moderate. This small performance advantage is usually outweighed by a tightening of manufacturing tolerances. Therefore, cone angles less than 30 degrees are generally not recommended for small standoff against surface targets. Ample experimental data show improved penetration at long standoffs for large cone angles; i.e., 80 to 120 degrees or more. The relationship between cone angle and penetration at long standoffs cannot be precisely predicted. However, it is generally accepted that, as the cone angle increases, standoff must also increase to achieve optimum penetration. Some investigators believe that this increase in standoff is a linear relationship, but see Figures 4-90 and 4-91a through 4-91d.

For use against aircraft at standoffs on the order of 100 feet, cone angles from 80° to 120° are recommended. (Reference 4-7.e.)

Cone thickness for best performance is primarily a function of cone apex angle and charge confinement, although other parameters play a lesser role. The optimum cone liner wall thickness increases with increasing cone angle and with increasing confinement of the charge. Generally, the optimum liner wall thickness varies between 2 and 4 per cent of the base diameter. However, some experiments indicate that thicknesses greater than this are acceptable. Work has been done using cone

Table 4-9
Relative Penetration Capabilities
of Various Liner Materials

Cone Material	Relative Penetration (In Mild Steel)
Aluminum	1.10
Copper	1.00
Steel	.75
Zinc	.65
Lead	.50
Glass	.40

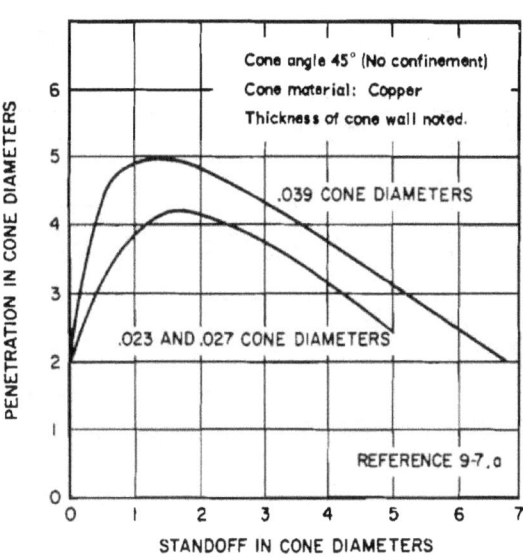

Figure 4-87. Penetration Vs. Standoff Against Mild Steel Target

thicknesses of 5, 6 and as high as 18 per cent. Thicknesses of about 6 per cent are generally used in warheads that are fired against aircraft at long standoffs.

Liner walls thinner than the optimum are characterized by excessive variation in penetration from round to round, and an overall decrease in penetration. Liner walls thicker than optimum also show a decrease in penetration although (1) it is slight for moderate increases in thickness and (2) the variation in round to round penetration is small. See Figures 4-92 and 4-93.

Explosive Charge Design. In general, penetration and hole volume increase with increasing charge length, and reach a maximum at about 2 or 2.5 charge diameters for heavily confined charges, and at about 4 charge diameters for lightly confined or unconfined charges. The usual effects of less than optimum charge length are lowered average penetration and reduced hole volume.

The explosive selection is important in shaped charge design. Explosives with high detonation pressure and velocity are the most desirable (see Appendix). The three explosives used most commonly in shaped charge ammunition are Composition B, Pentolite, and RDX. Other explosives which have been used to a lesser degree include 77/23:HBX/TNT, Composition A-3, and Composition C-3.

The explosive loading of the warhead must also be considered. The charge should be as homogeneous as possible, and should be free of accidental voids and foreign matter. It is necessary to incorporate any booster and wave shaping wells while casting or pouring the explosive. The loading in the region of the cone base is the most critical. The explosive in this region must be uniform in density and homogeneity.

The presence of voids in the charge is often responsible for loss in penetration. Despite the attempts usually made to exclude them, they are often present in some form. Sometimes the voids take the form of axially positioned pipes which are believed to have wave shaping effects. These axial pipes have been used to explain abnormally large penetrations sometimes encountered in test work. No known techniques have been developed for intentionally incorporating wave shaping voids in charges.

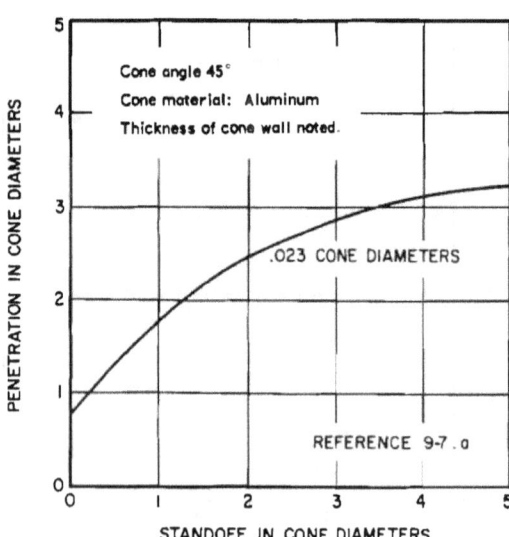

Figure 4-88. Penetration Vs. Standoff Against Mild Steel Target

Wave shaping offers one means of improving shaped charge performance. The purpose of the wave shaper is to invert the detonation wave and cause it to strike the cone wall at decreased angles of obliquity. All wave shapers are cylindrical and symmetrically placed between detonator and liner. Useful wave shaping might be accomplished by any of the following methods: inert fillers, other explosive fillers, voids in the explosive charge (pipes). (See Figure 4-94.) A warning should be given that, as of 1958, the application of wave shaping to shaped charges is still a difficult matter.

Inert solid cone-shaped fillers of glass or steel have produced a 20 per cent improvement in penetration performance without loss of hole volume. Cone shaped inert fillers with a base-to-altitude ratio of two have performed well, with little apparent degradation in performance for slightly different ratios. The base of the wave shaper is generally located immediately behind the apex of the cone, its diameter being only slightly less than that of the charge. Some inert fillers have taken the form of spherical segments. These have been positioned in the charge just forward of the booster, with the spherical surface toward the cone.

It is not necessary for the detonation wave to go around an inert filler in order to accomplish wave shaping. It is possible, with properly designed fillers in which the thickness and shape are adjusted, to allow the wave to pass right through the filler with suitable delay for wave refraction, and thus produce useful wave shaping that shows considerable improvement in penetration, without loss (or even with slight gain) in the hole volume. Explosive fillers that have been used for wave shaping include Baratol and TNT.

Peripheral initiation is one method of wave shaping which can be used to improve penetration. However, the actual improvement attained varies considerably with the liner material used. Also, hole volume may be increased by as much as 50 per cent by the use of peripheral initiation. When small asymmetries exist anywhere in the system, penetration will decrease. Although performance from carefully designed and accurately manufactured peripherally initiated rounds is superior to that obtainable from point-initiated rounds, the latter method of initiation yields more consistent results and also the point-inititated rounds are easier to manufacture.

Warhead Casing Design The case is designed to retain the charge before detonation and to confine the charge during detonation. The strength of the case required for confinement of the charge during detonation is practically nil in warheads where the length to diameter ratio of the charge exceeds about 4. As the L/D of the charge is reduced, the case strength required to confine the charge increases. Unfortunately, the casing thickness usually cannot be designed for optimum confinement. This is because in guided missile applications weight limitations will force the designer to use a lighter-than-optimum case while, for projectiles, set back forces will impose the use of a thicker-than-optimum casing.

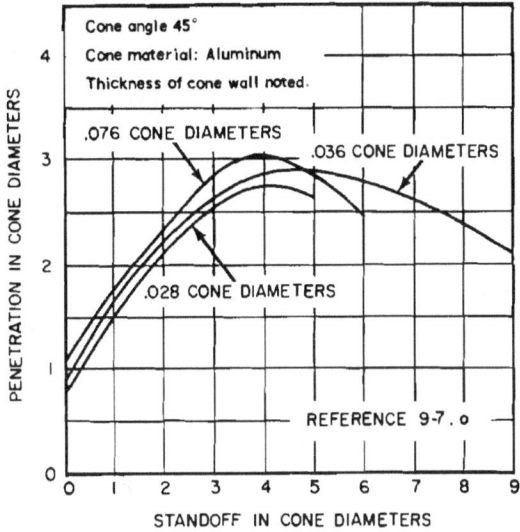

Figure 4-89. Penetration Vs. Standoff Against Mild Steel Target

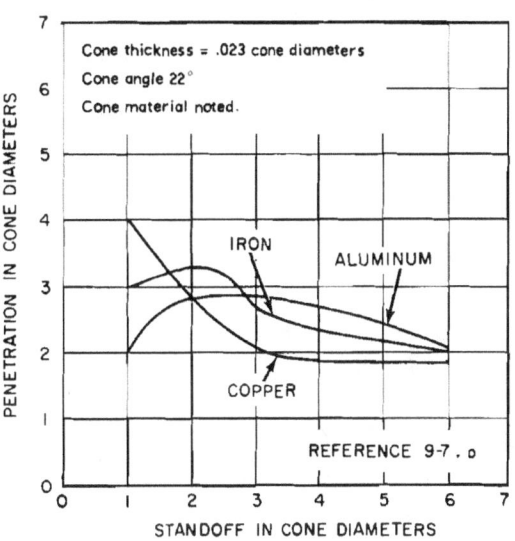

Figure 4-91a. Penetration Vs. Standoff Against Mild Steel Targets

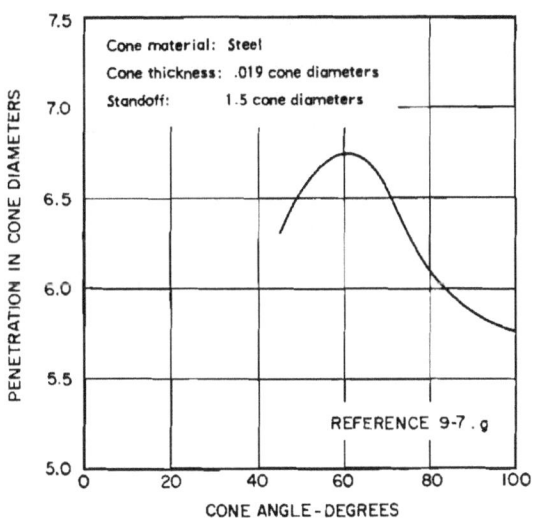

Figure 4-90. Penetration Vs. Cone Angle Against Concrete

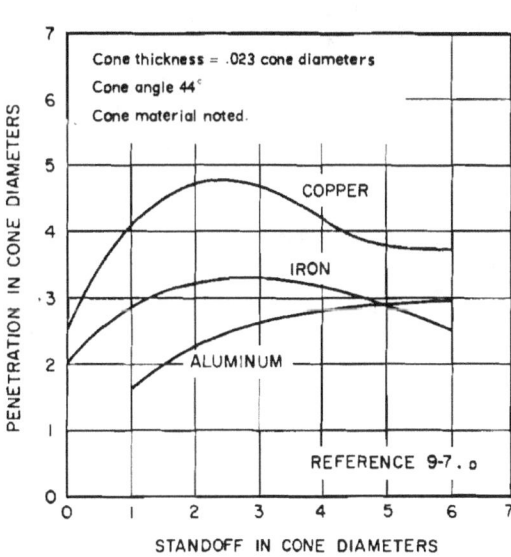

Figure 4-91b. Penetration Vs. Standoff Against Mild Steel Targets

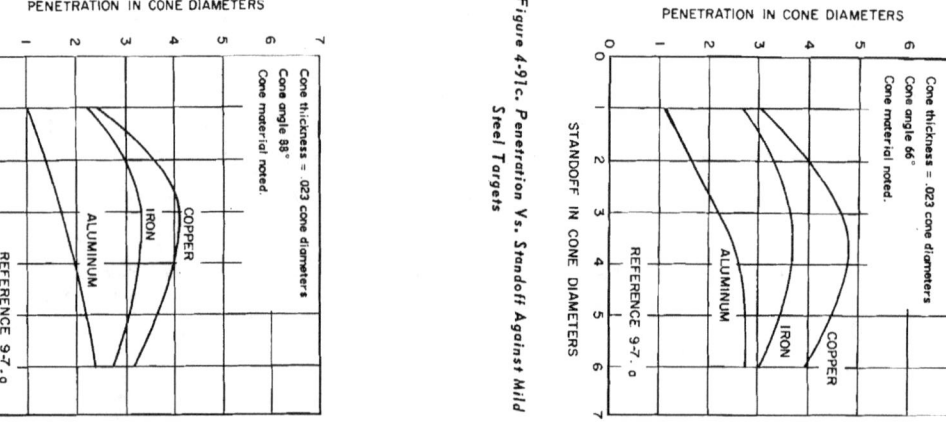

Figure 4-91c. Penetration Vs. Standoff Against Mild Steel Targets

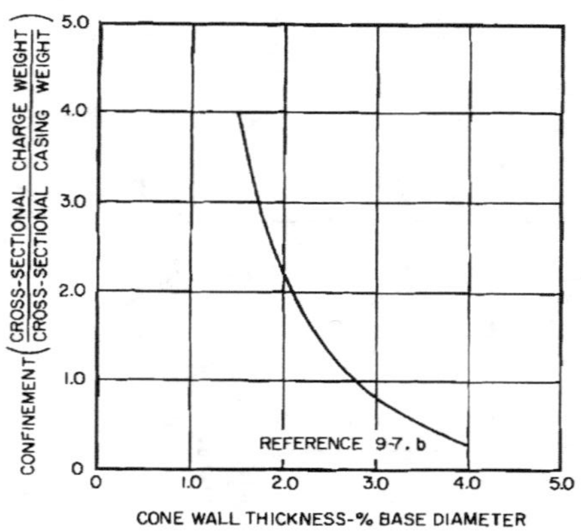

Figure 4-92. Reasonable Values of Cone Wall Thickness for Copper Cones (Apex Angles Between 40° - 45°)

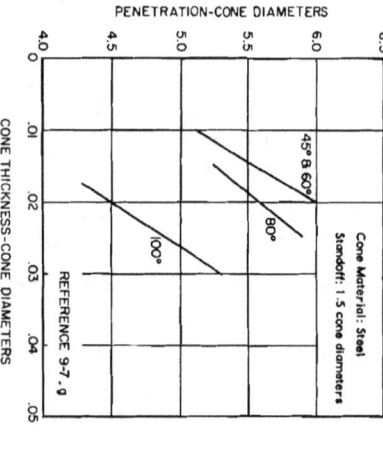

Figure 4-91d. Penetration Vs. Standoff Against Mild Steel Targets

Figure 4-93. Penetration Vs. Cone Thickness and Cone Angle Against Concrete

Summary of Fuzing Requirements The fuze designer needs design information to design a fuze which is compatible with the missile system and the warhead. He will have access to the same missile system data as does the warhead designer. In addition to this, the fuze designer will need the following information relating specifically to the warhead:
(1) Drawing of Warhead
(2) Standoff Distance
(3) Wave Shaping Used

Summary of Design Data At the conclusion of the design procedure, a summary of engineering data relating to the warhead should be prepared. This should include the following items:
(1) Total Weight
(2) Design and Installation Drawings
(3) Explosive
 (a) Material
 (b) Weight
 (c) Wave Shapers
 (d) Density
(4) Liner
 (a) Material—recommended manufacturing method
 (b) Thickness
 (c) Geometry
(5) Design Standoff Distance
(6) Center of Gravity Location
(7) Intended Performance

4-7.3. References

4-7.a "Critical Review of Shaped Charge Information", Edited by L. Zernow, BRL Report 905, ASTIA AD 48 311, dated May, 1954. "Liner Performance", John E. Shaw.

4-7.b "Critical Review of Shaped Charge Information", Edited by L. Zernow, BRL Report 905, ASTIA AD 48 311, dated May, 1954. "The Unfuzed Warhead", Hugh Winn.

4-7.c "A Theoretical Discussion of Penetration by Shaped Charge Jets with Some Experimental Results", J. B. Feldman, Jr., BRL "The Ordnance Corps Shaped Charge Research Report" No. 3-55, July, 1955.

4-7.d "The Effect of Shaped Charges at Long Standoff Against Aircraft", R. G. S. Sewell, L.

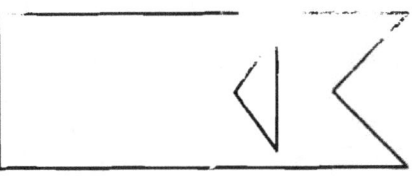
Cone Shaped Filler

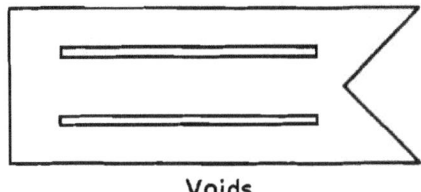
Voids

Spherical Segment

Peripheral Initiation

Explosive Insert

Figure 4-94. Explosive Charge Wave Shaping

N. Cosner, and J. Pearson, "The Ordnance Corps Shaped Charge Research Report", No. 4-55 ASTIA AD 83 088, October, 1955.

4-7.e "Studies of Damage to Aircraft Structures by Shaped Charges at Long Standoffs", NAVORD Report 2018, ASTIA AD 14162, March, 1953.

4-7.f "Manual for Shaped Charge Design", R. A. Brimmer, NAVORD Report 1248, August, 1950.

4-7.g "Penetration of Shaped Charge Jets into Concrete Targets", G. H. Jonas, BRL Tech. Note No. 939, September, 1954.

4-7.4. Bibliography

(1) "Collection and Arrangment of Shaped Charge Data", Arthur D. Little, Inc., Interim Report, October, 1955.

(2) "A Correlation of Explosive Properties with Shaped Charge Performance", NAVORD Report 2721, January, 1953.

(3) "Minutes of Shaped Charge Committee", Picatinny Arsenal, Shaped Charge Committee, February, 1954.

(4) "Multiple Fragment Effects in Shaped Charge Penetration", NOTS, BRL Report 837, November, 1951.

(5) "Test of Multiple Shaped Charge Terrier Warhead", University of Utah, Institute for the Study of Rate Processes, February, 1956.

(6) "Fundamentals of Shaped Charges", CIT, Final Report, October, 1954.

(7) "Shaped Charge Performance with Various Explosive Loadings", NAVORD Report 2767, February, 1953.

(8) "Preliminary Evaluation of Tests of the T-42 Type Shaped Charge Warhead", S. Wise, BRL Tech. Note 1079, June, 1956.

(9) "Some Remarks on the Performance of High Explosive Plastic Projectiles Against Armor Plate", F. I. Hill, BRL Memo. Report 519, December, 1950.

4-8. CHEMICAL AND BIOLOGICAL WARHEADS

4-8.1. Introduction Development, design, test, and evaluation work in the chemical and biological warhead fields is generally more highly classified than this pamphlet, and is very closely controlled and administered by the Army Chemical Corps, Edgewood, Maryland. Further, chemical and biological warheads are not designed as the primary warhead for a missile system, but rather are alternates. Thus the chemical and biological warhead designer will, in almost every case, find that the missile system data is firmly established and that the majority of the warhead detail design direction will be provided by personnel at Army Chemical Center, who will supply at least the following data:

(1) Number of Bomblets
(2) Bomblet Configuration and Weight
(3) Type of Bomblet Ejection System
(4) Bomblet Ejection Altitude
(5) Environmental and Storage Limitations
(6) Leak-Tightness Requirements
(7) Handling and Inspection Requirements
(8) Tactical Utilization

With this in mind, this subchapter of the pamphlet is written in narrative form rather than as a step-by-step design procedure. It is intended to acquaint the warhead designer with chemical and biological warhead design techniques in a general way.

4-8.2. Cluster-Type Warheads The cluster-type warhead is basically a container loaded with agent-filled bomblets which are randomly distributed over a given target area. It consists of a compartment filled with bomblets, a bomblet ejection system and fuze, and when necessary a means of maintaining the agent within specified environmental temperature limits. This type of warhead is generally located in the nose section of the missile, although occasionally the mid or after sections are used.

Bomblet Compartment and Structure The function of the warhead structure is dependent on its installation within the missile system. There are two basic types of warhead installation to be considered. The first of these is the case in which the warhead comprises a complete

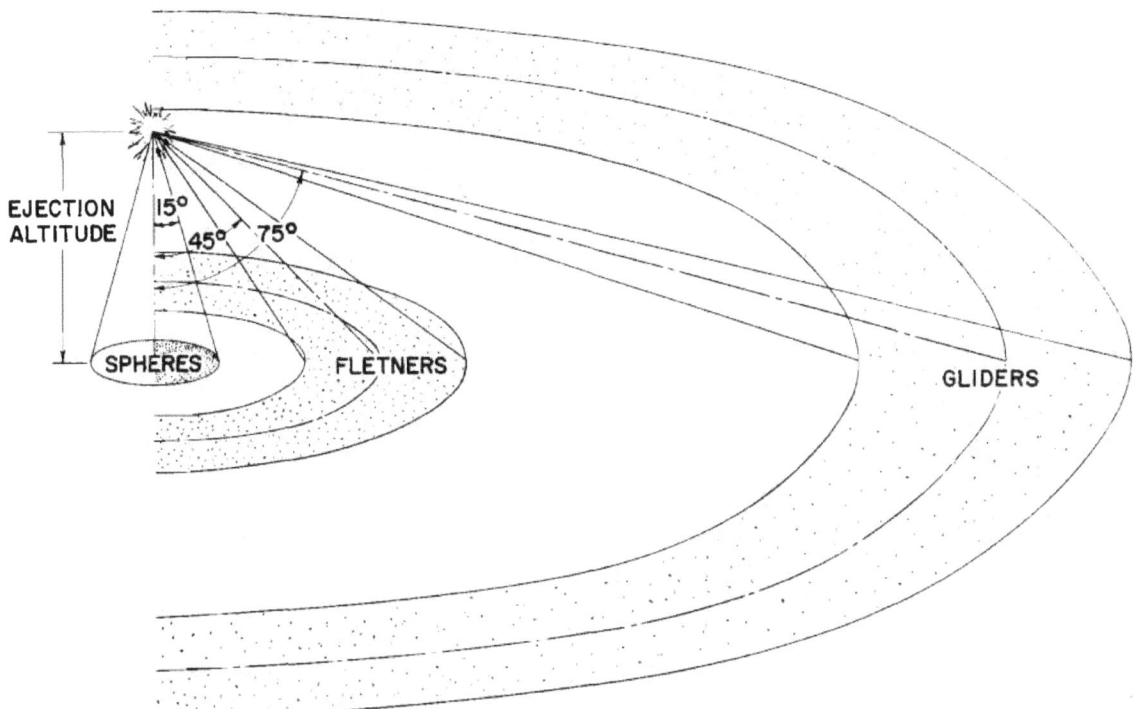

Figure 4-95. Bomblet Dispersion Patterns

compartment of the missile, terminated by a forward and rear bulkhead, and the second is the case where the warhead is installed within a compartment of the missile. The major design considerations on the former are that the warhead case is the aerodynamic skin of the missile, and that the warhead structure is an integral part of the missile structure and is usually detachable from the forward and rear missile components. For the latter installation, the missile skin must be removed to prevent interference with the warhead ejection mechanism. In either instance, both the installation and handling systems will influence the design and location of the main structural members of the warhead.

The bomblet ejection pattern from the warhead is intended to be equally distributed around the perimeter of the warhead. To accomplish this, the bomblet compartment or compartments are divided into longitudinal segments formed by placing longitudinal webs from the center of the warhead to the inner surface of the case. These webs are integrated in the warhead structure or they may be part of the bomblet ejection system. The bomblets are manually placed in these compartments according to loading patterns best suited to the bomblet and compartment configuration. The bomblets may be loaded into the warhead through the forward or rear bulkhead. When the forward or rear bulkheads cannot be removed, a transverse cut is made which divides the compartment into two halves. These two halves are then loaded separately and assembled.

Bomblets Early development of cluster-type warheads was directed toward utilizing the cylindrical type bomblets which had been developed for use in clustered bombs. The dispersion of these bomblets depended primarily on the variation of manufacturing tolerance, and the ground patterns obtained were too small to

Figure 4-96. Ejection Sequence, Spheres

fully utilize the agent contained. An attempt was then made to increase the dispersion by forceful ejection of the bomblets from the warhead. An increase in pattern size was realized but this was still unsatisfactory for warheads payloads of 1000 pounds or more. Recent development has been directed toward self-dispersing bomblets. Three basic configurations have been developed and are currently undergoing extensive testing. These configurations are a ribbed sphere approximately 4-1/2 inches in diameter, a Fletner, and a delta wing glider with a vertical fin. Examples of these shapes have been previously presented in Figure 1-34.

The development of these self-dispersing configurations has eliminated the need for a separate ejection system. The dispersion pattern obtained from the sphere is essentially circular, with random distribution of spheres throughout the circle as illustrated in Figure 4-95. The Fletners are randomly distributed throughout an annular ring as shown in Figure 4-95, and therefore must be ejected in groups of bomblets, each of which releases its individual units at a predetermined altitude in order to accomplish complete ground coverage. Initial investigation indicates that the delta wing gliders disperse themselves in a manner similar to the Fletner bomblets, but normally form a larger annulus as shown in Figure 4-95.

Ejection Systems The development of self-dispersing type bomblets has simplified the requirements of the ejection systems considerably. A direct result of this has been the elimination of the elaborate forceful ejection mechanism required on other bomblets. The self-dispersing type bomblets must be so removed from the warhead compartment as to clear the after-body of the missile in order to prevent damage to the bomblets.

The uniform distribution of bomblets within the impact pattern for the sphere configuration allows the release of the warhead bomblet load at one predetermined altitude. Because of structural and loading considerations longitudinal webs are sometimes inserted between longerons, 180° apart. A 120° cone is placed on the rear bulkhead. Detonating cord is inserted between the warhead skin and the longerons to provide for skin removal prior to the bomblet ejection. The forward and rear edge of the skin panels are not permanently fastened at the forward and rear bulkheads. In this type system the fuze ignites the detonating cord at the proper altitude, and the blast of the detonating cord and the air drag forces blow the skin panels away from the warhead structure. Aerodynamic forces then push the bomblets toward the rear bulkhead cone and the spheres slide off the cone and away from the missile structure. This operation is illustrated in Figure 4-96.

The annular distribution pattern for the Fletners and gliders requires ejection of groups of bomblets at more than one altitude to obtain complete coverage over the impact pattern as shown in Figure 4-97b. For purposes of illustration, the warhead is assumed divided into four longitudinal trapezoidal compartments as shown in Figure 4-97. These compartments are loaded with bomblets and attached to, and held in place by a gun tube which is an integral part of the missile structure. A high volumetric expansion, low pressure charge in the gun tubes ejects the compartments with sufficient velocity to clear the missile structure. The compartments incorporate a barometric fuze which ejects the contents of each compartment at a different altitude as shown in Figure 4-97. For each altitude at which a cluster of bomblets is opened, an annular pattern is produced on the ground as previously explained. The summation of these patterns gives the complete ground coverage shown in Figure 4-97. The largest annulus is obviously produced by the bomblets released at the highest altitude, and the maximum diameter of each annulus is proportional to the ejection altitude.

Environmental Requirements The warhead compartment environmental temperatures must be consistent with the specified bomblet environmental temperatures. It is therefore often necessary to supply some means of temperature control within the warhead. Insulation may be inserted between the warhead load and the warhead skin for this purpose. When an adequate amount of insulation cannot be installed within the warhead to maintain the required temperature, heating or cooling units must be used. A simple and efficient design for a heating unit utilizes the missile electrical power, and consists of electrically heated wire embedded in insulation which is sandwiched between two perforated sheets of light gauge metal. In a similar manner, cooling coils connected to an electrically operated refrigeration unit can be used. This type of unit can readily be preformed to the internal configuration of the warhead case. An additional advantage of these units is that they may be operated from an external power source when the missile electrical system is not in operation.

4-8.3. Massive-Type Warheads The massive-type warhead is basically a single container loaded with raw agent which is released immediately following impact with the target. It consists of the agent, container, and the agent dissemination mechanism and fuze.

This type of warhead is ordinarily limited to missile systems whose payload capacity is approximately 500 pounds or less. These warheads are capable of producing an extremely high agent concentration in the immediate vicinity of the impact area and are, therefore, well suited for use where relatively small concentrated type targets are the objective. The guidance accuracy required of the missile system for this type of warhead should be within the capabilities of guidance systems of present missiles in this payload range. The most critical design feature of the massive-type warhead is the leak-tightness required of the warhead case.

4-8.4. Agents The chemical agents are chemical substances whose toxic properties are such that they kill or incapacitate humans, domestic animals, or livestock through inhalation, ingestion, or absorption of agent through the skin. Some of these agents can kill or incapacitate within a very few minutes following exposure and are therefore well suited for tactical purposes. These agents are separated into physiological effect groupings of nerve, blister, blood, choking, tear and vomiting gas. Nerve gas is the most toxic of the above agents.

The biological agents are live disease organisms or their toxic products. These agents can kill or incapacitate humans, domestic animals, or livestock and can also destroy crops. The incubation time on biological agents varies from a few days to several months and, except for those agents with a very short incubation time, their employment is limited to strategic applications. Biological agents include fungi, bacteria, viruses, rickettsiae, protozoa, and toxins. For some of these agents, satisfactory immunization has not yet been developed.

Table 4-10 Characteristics of Existing Service Warheads

BLAST

General Type	Target	Total Weight lb	Length inches	Diameter inches	Status	Designation	c/m	Remarks	Missile
Surface-to-Surface	Light Structures and Material	3000 Nominal	72	30	Active	T3E3	3.46		Matador
Surface-to-Surface	Structures Susceptible to Blast	1500	43.0	29.47	Active	T2021	4.25		Honest John
Surface-to-Surface	General Ground Targets	1460 ±30	62.0	20.0	Inactive	T23E1	1.16	Has natural fragmentation	Corporal

CONTINUOUS ROD

General Type	Target	Total Weight lb	Length inches	Diameter inches	Status	Designation	c/m	Rod Number	Rod Size, inches	Expanded Ring Radius	Missile
Surface-to-Air	Aircraft	300	17.3	21	Developmental	Continuous Rod (Bomarc)	.800	800 approx.	3/16 x 1/4 x 16 1/2	100 feet	Bomarc
Surface-to-Air	Aircraft	405	21.9	23.3	Experimental	EX14 Mod 3	.673	534	0.250 x 0.250 x 20	125 feet	Talos (6b)
Surface-to-Air	Aircraft	180	20 max.	12 max.	Developmental	EX19	.722	274	1/4 x 1/4 x 18.3	65' (max.)	Advanced Terrier
Surface-to-Air	Aircraft	115	13.5	12 max.	Developmental	EX20 Mod 1	-	372	3/16 x 3/16 x length	55' radius 90° from W/H	Tartar
Air-to-Air	Aircraft	63	14.00	8.000	Developmental	EX21 Mod 1	.735	242	.187 x .187 x 10.3 (effective rod length 7.3)	27 feet (maximum theoretical)	Sparrow III

DISCRETE ROD

General Type	Target	Total Weight lb	Length inches	Diameter inches	Status	Designation	c/m	Rod Number	Rod Size, inches	Expanded Ring Radius	Missile
Air-to-Air	Aircraft	54	15	6.073 7.596	Experimental	EX1 Mod 0	-	120	0.45 x 0.33 x 3.875		Sparrow I
Air-to-Air	Aircraft	40	4	10	Inactive	145E	1.68	60	0.375 x 0.50 x 4		Oriole

Table 4-10 (Continued)

FRAGMENTATION

General Type	Target	Total Weight lb	Size Length inches	Size Diameter inches	Status	Designation	c/m	Fragments Number	Fragments Size, inches	Missile
Surface-to-Air	Aircraft	100	-	-	Developmental	XM-5	2.88	1800 approx.	1/2 x 1/2 x 1/4	Hawk
Air-to-Air	Aircraft	49	11.9	7.279 to 7.240	Developmental	EX 5 Mod 1	.741	1624	.375 x .375 x .3925	Sparrow II
Surface-to-Air	Aircraft	218	21.835	13.500 to 10.412	Active, Production	Mk 5 Mod 3	1.59	538 Size (a) 4058 Size (b)	3/8 x 3/8 x 3/4 3/8 cube	Terrier
Surface-to-Air	Aircraft	218	21.835	13.500 to 10.431	Active, Production	Mk 5 Mod 8	-	550 Size (a) 4200 Size (b)	3/8 x 3/8 x 3/4 3/8 cube	Terrier
Air-to-Air	Aircraft	44 ± 1 (without fuzing)	15	7-5/8 to 8-1/8	Production	Mk 7 Mod 0	.944	1315 approx.	0.312 x 0.401 x 0.4	Sparrow I
Air-to-Air	Aircraft	25	13-1/2	5	Active	Mk 8 Mod 0	1.82	1300 approx.	-	Sidewinder
Surface-to-Surface	Personnel	1345 ± 20 (without fuzing)	33.79 max.	22.45 max.	Interim	T25E1	.181	58000 approx.	.150 x .150 x 2.70	Corporal
Surface-to-Air	Aircraft	11.75 ± .25	6.393 max.	5.16 max.	Active	T28E4	-	800 approx.	-	Nike 1
Surface-to-Air	Aircraft	176.75 ± .25	21.36	11.7	Active	T37E3	1.19	8815	.3125 ± .003 cube	Nike 1
Surface-to-Air	Aircraft	121.25 ± .25	21.4	11.0	Active	T38E3	.698	4418	.3125 ± .003 cube	Nike I
Surface-to-Surface	Personnel	1345 ± 20 (without fuzing)	-	-	Developmental	T40	.083	500,000	.090 dia. x 1.25 lg	Corporal
Surface-to-Air	Aircraft	1118 max. 1096 min.	27.10	29.24	Developmental	T45	-	18,902	.414 cube	Nike Hercules
Air-to-Air	Aircraft	63	14.000	8.000	Superseded by EX21	EX2 Mod 3	.542	1488	.375 x .375 x .516	Sparrow III
Surface-to-Air	Aircraft	350	15.920	26.280 to 23.840	Inactive	EX7 Mod 1	2.18	8200	3/8 cube	Talos (6a)
Surface-to-Air	Aircraft	145.76	-	-	Superseded	T22E4	1.94	9140 (design) 9117 (on test model)	.2525 cube	Nike I
Surface-to-Air	Aircraft	405	21.920	28.3 to 23.8	Superseded by Continuous Rod Warhead	EX17 Mod 1	2.20	7650	3/8 cube	Talos (6b)
Air-to-Air	Aircraft	A = 31.07 B = 28.92 C = 35.70 D = 41.24	4	10	Inactive	145 A, B, C, D	A = 1.03 B = 0.74 C = 1.33 D = 1.87	Natural Fragmentation	Natural Fragmentation	Oriole

Table 4-10 (Continued)

CLUSTER

General Type	Target	Total Weight lb	Size Length inches	Size Diameter inches	Status	Designation	Submissile Type	Number	Weight of Submissile	Weight Charge per Submissile	Missile
Surface-to-Air	Aircraft Bomber	300	10.4	32.2	Developmental	Cluster Warhead (Bomarc)	Unstabilized	48	4.2 (approx.) including fuze	2.9	Bomarc
Surface-to-Air	Aircraft	156.14	-	-	Developmental	Cluster Warhead (Nike I) Center Cluster	Unstabilized	28	4.56	2.9	Nike I
Surface-to-Air	Aircraft	127.04	-	-	Developmental	Cluster Warhead (Nike I) After Cluster	Unstabilized	21	4.56	2.9	Nike I
Surface-to-Air	Aircraft	-	-	-	Developmental	T-46	Unstabilized	-	4.42	-	Nike Hercules
Air-to-Air	Aircraft	60.39 to 57.89	26.25	-	Experimental	Dispersal Warhead (Sparrow I)	Fin Stabilized	16	-	1.4	Sparrow I

SHAPED CHARGE

General Type	Target	Total Weight lb	Size Length inches	Size Diameter inches	Status	Designation	Cone Liner Apex Angle Degrees	Cone Liner Wall Thickness inches	Material	Base Diameter inches	Missile
Surface-to-Surface	Tanks and Armored Vehicles	10.60	15.15	-	-	French SS10	-	-	Copper	6.48	French SS10
Surface-to-Surface	Bunkers and Fortifications	500 approx.	81.2	20.5 max.	Developmental	T34	40°	0.25	Copper	14.25	Lacrosse

Figure 4-97. Ejection Sequence, Fletners and Gliders

4-8.5. Bibliography

(1) "Chemical and Biological Agents Chart for Guided Missiles", ATI No. 455, July, 1946.

(2) "Development of Special Chemical Heads for Rockets", Arthur D. Little, Inc., ATI No. 131213, January, 1952.

(3) "Launching of HVAR Heads", Hughes Aircraft Co., ATI No. 136846 and 136847, April and May, 1950.

4-9. CHARACTERISTICS OF SERVICE WARHEADS

Pertinent Information relative to existing service warheads for guided missiles is included in this subchapter for reference purposes. It was obtained from "Survey of Guided Missile Warheads", Haller, Raymond and Brown, Inc., September, 1956, Report No. 91-R-5. It is to be noted that many additional service warheads have been developed or put in production since the publication of the above Survey.

Chapter 5
WARHEAD EVALUATION

5-1. EVALUATION PRINCIPLES

5-1.1. Introduction Whenever any warhead has been developed, it is necessary to know the adequacy of this warhead relative to the original requirement to which it was developed; that is, final evaluation of the warhead must be made. In order to have this evaluation be more than an opinion, it must be based on factual data as nearly as possible.

A reliable evaluation of a specific warhead against a specific target can be obtained by conducting an adequate number of tests where the warhead is employed in the given guided missile system against the target and under conditions closely approximating actual engagement conditions. Such tests, carried out using remotely controlled or stationary targets are generally useful, reliable and sufficiently accurate, although expensive. Although the evaluating of a warhead in a complete missile system is a highly complex problem, the warhead designer can, without conducting these elaborate tests, arrive at approximate evaluations which are usable for comparative purposes. Such approximate evaluations give the warhead designer an insight into the relative efficiency of proposed designs as well as a method of ranking these proposed designs in order of their adequacy of meeting the original requirements for which they were developed.

There are several ways to present the evaluations of specific warheads employed against specific targets. Chief among these and the one treated in this pamphlet is that of giving probabilities of inflicting specified kinds of damage (called "kills") upon the specified targets under specified conditions by specified warheads.

Except for evaluation techniques which involve the firing of actual warheads against actual targets, and the subsequent assessment of the damage inflicted thereto, the methods used and discussed in this chapter are based on mathematical analyses. The principles involved are mentioned, as well as some specific nomenclature, etc., but it is not the purpose of this pamphlet to provide the reader with a basic knowledge of the theory of probability. For the latter, the reader is directed to standard texts in the field such as those listed under the Bibliography, 5-6, entries 13, 14, and 15.

5-1.2. Overall Kill Probability The overall probability of inflicting a specified kind of damage (called "overall kill probability") upon a specified target under specified conditions by a specified missile system having a specified warhead is equal to the product of the following probabilities:

P_r = the probability of detecting and/or recognizing the target

P_c = the probability that the missile system will launch the missile (conversion)

P_d = the probability that the missile will deliver the warhead to the target

P_f = the probability that the fuzing system will function and that the warhead will have a high order detonation

P_k = the probability of inflicting the specified damage (kill) provided the target is detected (and/or recognized), the missile system functions, the warhead is delivered to the target, and the fuzing system functions (called "conditional kill probability")

Thus, if P_s = overall kill probability (success), then

$$P_s = P_r \cdot P_c \cdot P_d \cdot P_f \cdot P_k, \quad (5\text{-}1.1)$$

This overall kill probability is of interest to the warhead designer in that he should know how his design fits into the overall missile system. However, the warhead designer has no control over recognition (detection), conversion, and delivery. He has little control over the fuzing reliability.

It is to be noted that from an economic or supply standpoint the "expected fraction killed" of an area target with a given number of shots is of interest in order to determine the weapon to be used. However, in the battlefield, of primary concern is the high probability of obtaining a given level of kill; e.g., 90% confidence of killing at least 50% of an area target. It is to be further noted that improvements in reliability of guided missiles are generally of more importance than increases in their single shot kill probability P_k.

As can be seen by equation (5-1.1), the warhead designer can contribute to the overall kill probability, P_s, by maximizing as much as possible the conditional kill probability, P_k, over which he does have control within the limits of the parameters given him. Thus, if P_{k_i} is the conditional kill probability for the i'th warhead design, the maximum P_{k_i} (i.e. P_{k_i} max) would be the criterion for use in determining the warhead design which would be most adequate relative to the original requirements for which the warheads were developed.

5-1.3. Conditional Kill Probability The conditional kill probability, P_k, which is of concern to the warhead designer is that probability of inflicting the specified damage (kill) provided the target is detected (and/or recognized), the missile system functions, the warhead is delivered to the target, and the fuzing system functions. This conditional kill probability is the means whereby the warhead designer may evaluate the relative effectiveness of proposed warhead designs and rank these designs in order of their adequacy of meeting the original requirements for which they were developed. In terms of effectiveness the warhead with the maximum conditional kill probability, P_{k_i} max, is the best.

The conditional kill probability, P_k, of a warhead is a function of the following:

$\theta(G)$ = frequency distribution of the guidance error

$\Psi(F)$ = frequency distribution of the fuzing error

V_m = velocity of the missile which carried the warhead, measured in the direction of travel of the missile in feet per second

V_t = velocity of the target measured in the direction of travel of the target in feet per second

θ = angle between the missile trajectory and the target trajectory

h = altitude of engagement

$l(m)$ = lethality of the missile warhead

$V(T)$ = vulnerability of the target.

Thus, (5-1.2)
$$P_k = f\left[\phi(G), \Psi(F), V_m, V_t, \theta, h, l(m), V(T)\right]$$

The frequency distribution of the guidance error, $\phi(G)$, is discussed in Section 5-1.4.

The frequency distribution of the fuzing error, $\Psi(F)$, is discussed in Section 5-1.5.

The velocity of the missile, V_m, the velocity of the target, V_t, and the altitude of engagement, h, are normally specified and may be treated as constants. However, the warhead designer has the responsibility of recommending that a missile with inadequate payload or speed advantage over the target be superseded.

The angle, θ, between the missile and target trajectories is generally a variable that must be considered in the evaluation of a warhead. However, θ can be held constant so that several warheads may be evaluated and compared at a specific engagement aspect.

The lethality of the missile, $l(m)$, is a function of the warhead type and of the variables over which the warhead designer does have control within the design parameters. For antiaircraft warheads, the speed ratio V_m/V_t is extremely important. These variables are treated in the individual warhead design sections.

The vulnerability of the target, $V(T)$, depends upon its shape, size, location, structure, toughness, motive power, maneuverability, payload, special distinguishing characteristics, attack time, whether it is singly or multiply vulnerable to the warhead considered, and other descriptive data.

5-1.4. Distribution of Guidance Error Guidance error is defined as the perpendicular distance from the aim point to the missile trajectory. The aim point is defined as that point where the missile warhead would detonate if the guidance system and the fuzing system were to function in an ideal manner. Because of human and mechanical factors, neither guidance nor fuzing function perfectly and therefore guidance and fuzing error have frequency distributions. Assuming there is no overall bias, then these errors are taken to be distributed around the aim point.

In order to gain an insight into the distributions of these errors, one must first become oriented to the velocity vector relationship of the missile and the target. This can be approached by first looking at the special case where the target is an aircraft and the aim point is the geometrical center of the aircraft. Thus, with perfect guidance and perfect fuzing the attack would occur as shown in Figure 5-1.

Figure 5-1 shows the relative positions of the target and missile at some specific time just prior to the time of impact. In this picture it can be seen that the guidance system has accounted for the velocity-vector relationship of the missile and target so that the aim-point is that point where the geometrical center of the target was expected to be at the instant the missile intersected the path (trajectory) of the target center.

Now, if a plane normal to the missile trajectory is drawn through the aim point, the x-axis of this plane is in the direction of the yaw of the missile and the y-axis is in the direction of pitch of the missile, both with origin at the aim-point. This is illustrated in Figure 5-2.

Figure 5-2 also shows a z-axis originating at the aim-point and normal to the x,y plane. Fuzing error is distributed along the missile trajectory which is along or parallel to the z-axis. The distribution of fuzing error is treated in Section 5-1.5. Assuming that fuzing error (i.e. error in the z direction) is independent of guidance error (i.e. error in the x, y plane), and analysis of the distribution of guidance error is reduced to a two-dimension problem.

Now, if a great number of missiles were fired at the aim-point, due to random errors in the guidance system (human, mechanical, and electronic) the missile trajectories could intersect the x, y plane at any points as shown in Figure 5-3. If the frequency distribution of these points were known, one could calculate the probability that a single missile will intersect the x,y plane within any defined area.

As can be seen in Figure 5-3, the guidance error of the i'th missile is Δ_i, and Δ_i has a component on both axes, x_i and y_i, where

$$\Delta_i = \sqrt{x_i^2 + y_i^2} \qquad (5-1.3)$$

It is assumed that the x and y are independent. Thus the frequency distributions of x and y may be analyzed independently. Consider first the frequency distribution of x. It is generally assumed that guidance errors are random occurences and that the axis component of these errors follow the so-called normal (Gaussian) curve of error. In this case the frequency distribution of x is mathematically defined by

$$\phi(x) = \frac{1}{\sigma_x \sqrt{2\pi}} e^{-\frac{x^2}{2\sigma_x^2}} \quad (5-1.4)$$

where σ_x = standard deviation of the x from the origin (aim-point).

The frequency function $\phi(x)$ can be thought of as a distribution of ratio of occurrence of each x_i when a very large number of missiles is fired. If N is the total number of missiles fired and n_{xi} is the number of these missiles whose x component of error lies between x_i and $x_i + \Delta x_i$, then, as $\Delta x_i \to 0$,

$$\phi(x_i) \to \frac{n_{xi}}{N}. \quad (5-1.5)$$

A graph of the normal frequency function, $\phi(x)$, is given in Figure 5-4.

The probability that any random occurrence of x is between any two given values, say $\pm \sigma_x$, is the area between the normal frequency curve and the x-axis. In this case the probability is equal to .68 and is illustrated in Figure 5-5.

The standard deviation, σ_x, is a function of the guidance system over which the warhead designer has little or no control. Therefore, it is assumed here that this value is given to him by the missile system designer. Thus, if it is known that a specified target will be given a specified damage when $x = \pm 30$ feet and the given $\sigma_x = 30$ feet, then the probability that a random shot with the specified warhead will inflict the specified damage on the specified target is .68, neglecting possible errors in the y and z coordinates.

Let $t = \frac{x}{\sigma_x}$ \quad (5-1.6)

Then the normal frequency function $\phi(t)$ is

$$\phi(t) = \frac{1}{\sqrt{2\pi}} e^{-t^2/2} = \sigma_x \phi(x) \quad (5-1.7)$$

and is illustrated in Figure 5-6.

A table of the areas between $-t$ and t for .01 steps of the deviate t is given in Table 5-1, for the convenience of the warhead designer in evaluating proposed designs. These areas may also be found in any standard book of mathematical tables under the title of "Areas of the Normal Curve of Error". (Reference 5-b)

Since y is an independent variable and assumed to be normally distributed, its frequency distribution function is

$$\phi(y) = \frac{1}{\sigma_y \sqrt{2\pi}} e^{-y^2/2\sigma_y^2} \quad (5-1.8)$$

where σ_y = standard deviation of the y from

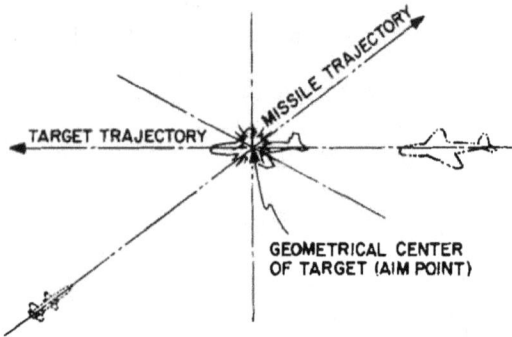

Figure 5-1. Attack with Perfect Guidance and Perfect Fuzing

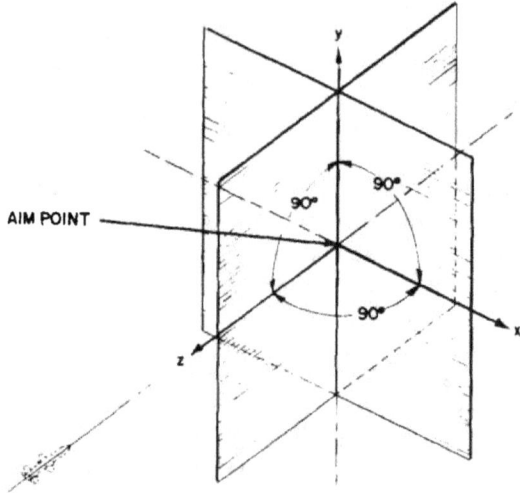

Figure 5-2. Orientation of the Axes

the origin (aim-point). The reader should note the similarity of this function to equation 5-1.4. Thus, if

$$t = \frac{y}{\sigma_y} \quad (5-1.9)$$

then the normal frequency function is identical to equation 5-1.7, i.e.,

$$\phi(t) = \frac{1}{\sqrt{2\pi}} e^{-t^2/2} = \sigma_y \phi(y) \quad (5-1.10)$$

and Table 5-1 may be used to find the probability that a random shot will give a y_i between any specified y (plus and minus). Once again, σ_y is to be considered as being given.

Probability of Hitting a Rectangular Area If fuzing is perfect, if the target is a known rectangular area, and if there is a known probability of kill provided the area is hit, P_c, then the conditional kill probability, P_k, is the product of P_c and the probability of a hit, P_h on the area, i.e.,

$$P_k = P_c \cdot P_h \quad (5-1.11)$$

Thus, to find P_k, the problem is reduced to finding, P_h, the probability of hitting the area and multiplying this by P_c.

P_h can be thought of as the ratio of the number of shots which hit within the perimeter of the area to the total number. If this area is oriented as shown in Figure 5-7 so that it is a $2a \times 2b$ area with the x and y axes oriented at the center, P_h is the product of P_a the probability that the x component of a hit is between $\pm a$, and P_b, the probability that the y component of a hit is between $\pm b$, i.e.,

$$P_h = P_a \cdot P_b \quad (5-1.12)$$

Assuming known and independent normal distributions in the directions of both axes, then σ_x and σ_y are known. P_a is the area of the normal curve of area and can be found by letting

$$t = \frac{a}{\sigma_x}, \quad \text{and } P_a = \frac{1}{\sqrt{2\pi}} \int_{-t}^{t} e^{\frac{-t^2}{2}} dt \quad (5-1.13)$$

and finding the area. Table 5-1.

In a similar manner, P_b can be found by letting

$$t = \frac{b}{\sigma_y} \quad (5-1.14)$$

and finding the area from Table 5-1.

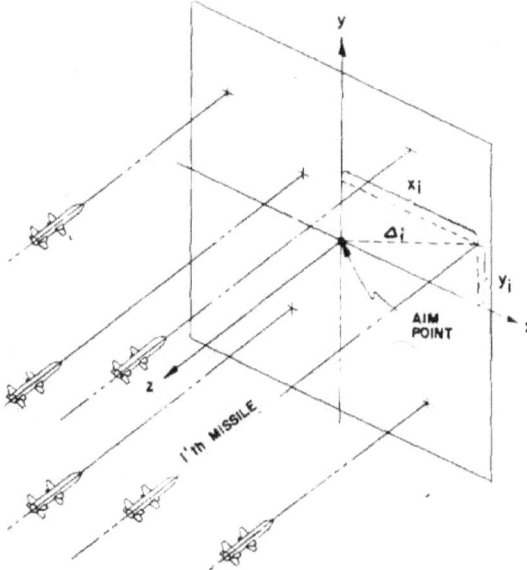

Figure 5-3. Random Guidance Errors

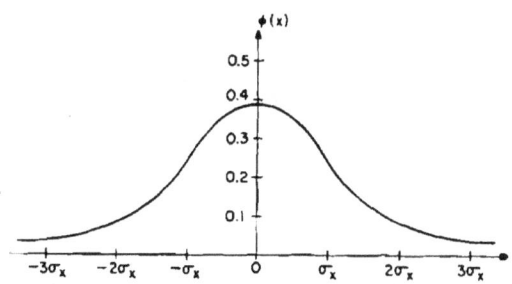

Figure 5-4. The Normal Frequency Function

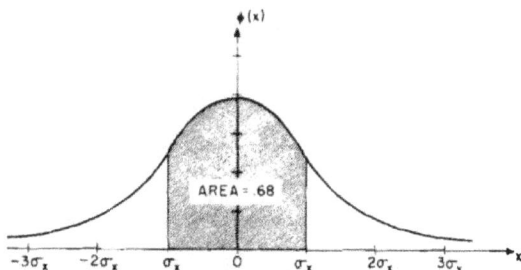

Figure 5-5. Area Under Normal Frequency Curve

Thus, if fuzing is perfect and P_c, σ_x and σ_y are known for a given rectangular target 2a x 2b, then the conditional kill probability is

$$P_k = P_c \cdot P_a \cdot P_b \qquad (5\text{-}1.15)$$

Circular Normal Distribution With normal distributions in x and y and $\sigma_x = \sigma_y$, the probability that a missile will hit the x, y plane within a circle of radius, Δ, from the aim-point is

$$P_\Delta = 1 - e^{-\frac{\Delta^2}{2\sigma_G^2}} \qquad (5\text{-}1.16)$$

where σ_G = linear standard error of guidance and determines the frequency distribution of Δ. In this case $\sigma_G = \sigma_x = \sigma_y$.
Letting,

$$u = \frac{\Delta}{\sigma_G} \qquad (5\text{-}1.17)$$

the probability of a hit within a circle of a radius of u standard errors of guidance is

$$P_\Delta = 1 - e^{-\frac{u^2}{2}} \qquad (5\text{-}1.18)$$

A graph of P_Δ is shown in Figure 5-8 and a table of values of P_Δ for .01 steps of the deviate u is given in Table 5-2.

Elliptical Normal Distribution With normal distributions in x and y and $\sigma_x \geq \sigma_y$, then P_R, the probability that a missile will hit the x-y plane within a circle of radius R can be read directly from Table 5-3 and Figure 5-9 for values of $\sigma_y / \sigma_x = \sigma_{min} / \sigma_{max}$, and R/σ_{max}. The tabulated values were furnished by the Computing Laboratory of the Ballistic Research Laboratories.

The phrase "Elliptical Normal Distribution" is generally used to describe the distribution of the radial density function in a two-dimensional, non-circular, normal distribution.

Miss Distance One definition of miss distance is the shortest distance between the center of the warhead burst and the geometrical center of the target. The miss distance is a combination of guidance error and fuzing error.

In many instances the aim-point is other than the geometrical center of the target. Each particular combination of missile, guidance system, warhead, and target would have its own unique solution relating miss distance to guidance error. This solution can be found through analytic geometry. No standard deviation of miss distance is a valid estimate of σ_G, the standard error of guidance, unless it has first been subjected to this analysis.

In addition, due to the limitations of the capabilities of electronics measuring devices, in many cases the so-called "miss distance" tabulated from test missile firings against drones is not the same as the miss distance defined previously. Instead, this "miss distance" is the shortest distance between the missile trajectory and the perimeter of the target if the measuring device is mounted in the missile, or between the device's antenna

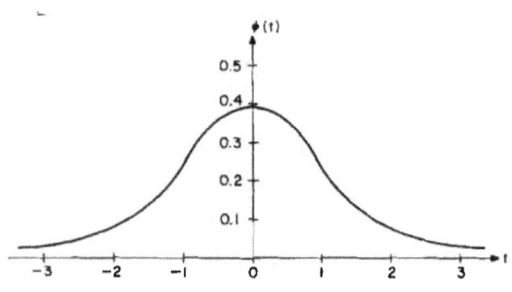

Figure 5-6. The Normal Curve of Error

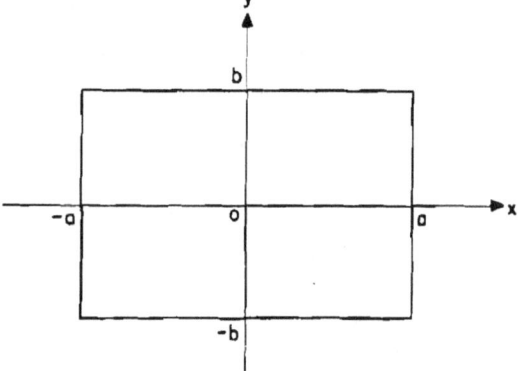

Figure 5-7. Orientation of Rectangular Target Area

Table 5-1 The Areas Under The Normal Curve Of Error (Included Between t and -t)

t	.00	.01	.02	.03	.04	.05	.06	.07	.08	.09
.00	.000	.008	.016	.024	.032	.040	.048	.056	.064	.072
.10	.080	.088	.096	.103	.111	.119	.127	.135	.143	.151
.20	.159	.166	.174	.182	.190	.197	.205	.213	.221	.228
.30	.236	.243	.251	.259	.266	.274	.281	.289	.296	.303
.40	.311	.318	.326	.333	.340	.347	.354	.362	.369	.376
.50	.383	.390	.397	.404	.411	.418	.425	.431	.438	.445
.60	.451	.458	.465	.471	.478	.484	.491	.497	.503	.510
.70	.516	.522	.528	.535	.541	.547	.553	.559	.565	.570
.80	.576	.582	.588	.593	.599	.605	.610	.616	.621	.627
.90	.632	.637	.642	.648	.653	.658	.663	.668	.673	.678
1.00	.683	.688	.692	.697	.702	.706	.711	.715	.720	.724
1.10	.729	.733	.737	.742	.746	.750	.754	.758	.762	.766
1.20	.770	.774	.778	.781	.785	.789	.792	.796	.799	.803
1.30	.806	.810	.813	.816	.820	.823	.826	.829	.832	.835
1.40	.838	.841	.844	.847	.850	.853	.856	.858	.861	.864
1.50	.866	.869	.871	.874	.876	.879	.881	.884	.886	.888
1.60	.890	.893	.895	.897	.899	.901	.903	.905	.907	.909
1.70	.911	.913	.915	.916	.918	.920	.922	.923	.925	.927
1.80	.928	.930	.931	.933	.934	.936	.937	.939	.940	.941
1.90	.943	.944	.945	.946	.948	.949	.950	.951	.952	.953
2.00	.954	.956	.957	.958	.959	.960	.961	.962	.962	.963
2.10	.964	.965	.966	.967	.968	.968	.969	.970	.971	.971
2.20	.972	.973	.974	.974	.975	.976	.976	.977	.977	.978
2.30	.979	.979	.980	.980	.981	.981	.982	.982	.983	.983
2.40	.984	.984	.984	.985	.985	.986	.986	.986	.987	.987
2.50	.988									
2.60	.991									
2.70	.993									
2.80	.995									
2.90	.996									
3.00	.997									

location and the perimeter of the missile if the device is located in the target. It is readily seen that this measured distance is frequently less than the defined miss distance by a figure between the minimum and the maximum distances between the c.g. and the target perimeter. With a large target such as a bomber this difference could be a hundred or more feet and would be highly significant. At best, the measured "miss distance" requires conversion before it could be considered or used where the defined miss distance is called for. Because of these shortcomings, difficulties, and ambiguities, the term "miss distance" has purposely been avoided in this pamphlet.

5-1.5. Distribution of Fuzing Error Fuzing error is defined as the shortest distance from a plane normal (perpendicular) to the missile trajectory and passing through the aim-point to the point of actual detonation of the warhead. The aim-point, as defined previously, is that point where the missile warhead would burst if there were both normal guidance and fuzing without bias.

As explained in Section 5-1.4, human, mechanical and electronic factors cause the fuze to function imperfectly and result in fuzing error. Assuming no bias in fuzing, these errors are distributed around the point where the x-y plane of Section 5-1.4 intersects the missile trajectory and along the missile trajectory. These errors would then have z components. Thus, fuzing error can be considered as being distributed along the z-axis and around the aim-point. Figure 5-10 shows random fuzing errors in a missile with random guidance errors. As shown, the z_i may occur on either side of the x-y plane.

If the frequency distribution of these errors were known, one could calculate the probability that a single fuze will detonate the warhead within any given distance from the x-y plane.

It is generally assumed that fuzing errors are random occurrences and that they follow

Table 5-2 The Probabilities of a Hit, P_m, Within a Circle of Radius u Standard Errors

u	.00	.01	.02	.03	.04	.05	.06	.07	.08	.09
0.00	.000	.000	.000	.000	.001	.001	.002	.002	.003	.004
0.10	.005	.006	.007	.008	.010	.011	.013	.014	.016	.018
0.20	.020	.022	.024	.026	.028	.031	.033	.036	.039	.041
0.30	.044	.047	.050	.053	.056	.059	.063	.066	.070	.073
0.40	.077	.081	.085	.088	.092	.096	.100	.105	.109	.113
0.50	.117	.122	.127	.131	.136	.140	.145	.150	.155	.160
0.60	.165	.170	.175	.180	.185	.190	.196	.201	.206	.212
0.70	.217	.222	.228	.234	.239	.245	.251	.257	.263	.268
0.80	.274	.280	.286	.292	.297	.303	.309	.315	.321	.327
0.90	.333	.339	.345	.351	.357	.363	.369	.375	.381	.388
1.00	.393	.399	.406	.412	.418	.424	.430	.436	.441	.448
1.10	.454	.460	.466	.472	.478	.484	.490	.496	.502	.507
1.20	.513	.519	.525	.531	.537	.542	.548	.554	.560	.565
1.30	.571	.576	.582	.587	.593	.598	.603	.609	.614	.620
1.40	.624	.630	.635	.640	.645	.650	.655	.661	.666	.671
1.50	.675	.680	.685	.690	.694	.699	.704	.708	.713	.718
1.60	.722	.726	.731	.735	.739	.744	.748	.752	.756	.760
1.70	.764	.768	.772	.776	.780	.784	.788	.791	.795	.798
1.80	.802	.806	.809	.813	.816	.819	.823	.826	.829	.832
1.90	.836	.839	.842	.845	.840	.851	.854	.856	.859	.862
2.00	.865	.868	.870	.873	.875	.878	.880	.883	.885	.888
2.10	.890	.892	.894	.896	.899	.901	.903	.905	.907	.909
2.20	.911	.913	.915	.917	.919	.921	.922	.924	.926	.928
2.30	.929	.931	.932	.934	.935	.937	.938	.940	.941	.943
2.40	.944	.945	.947	.948	.949	.950	.951	.953	.954	.955
2.50	.956	.957	.958	.959	.960	.961	.962	.963	.964	.965
2.60	.966	.967	.968	.969	.969	.970	.971	.972	.973	.973
2.70	.974	.975	.975	.976	.977	.977	.978	.979	.979	.980
2.80	.980	.981	.981	.982	.982	.983	.983	.984	.984	.985
2.90	.985	.986	.986	.986	.987	.987	.988	.988	.988	.989
3.00	.989									

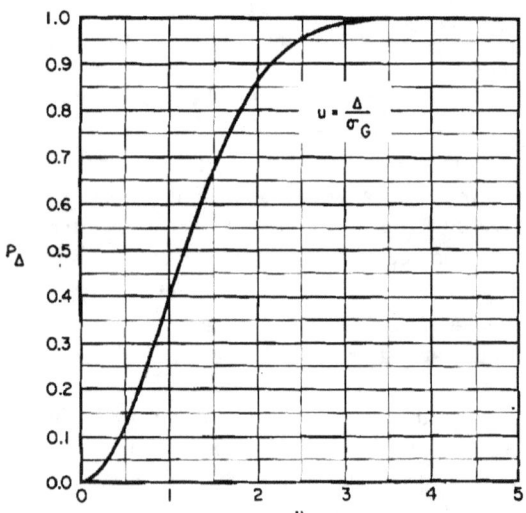

Figure 5-8. Probability of a Hit, P_Δ within a Circle of u Standard Errors

the normal curve of error. In this case the frequency distribution of the z, $\phi(z)$, is defined by

$$\phi(z) = \frac{1}{\sigma_z \sqrt{2\pi}} e^{-\frac{z^2}{2\sigma_z^2}} \quad (5-1.19)$$

where σ_z = standard error of fuzing.

For a discussion of this distribution, see Section 5-1.4 and note the similarity between equation 5-1.19 and equation 5-1.4. Now let

$$t = \frac{z}{\sigma_z} \quad (5-1.20)$$

Then the normal frequency function is

$$\phi(t) = \frac{1}{\sqrt{2\pi}} e^{\frac{-t^2}{2}} \quad (5-1.21)$$

and this frequency function is illustrated in Figure 5-6.

The areas between $-t$ and t for .01 steps of the deviate t given in Table 5-1 can be used in finding P_z, the probability of fuzing within any given z of the x-y plane. That is,

$$P_z = \phi(t) = \frac{1}{\sqrt{2\pi}} \int_{-t}^{t} e^{\frac{-t^2}{2}} dt \quad (5-1.22)$$

5-1.6. Damage Classification The purpose of any missile system is to help prevent enemy use

Table 5-3 Cumulative Bivariate Normal Distribution Over Circles Centered At The Mean

R/σ MAX. \ σ MIN./σ MAX.	0	.1	.2	.3	.4	.5	.6	.7	.8	.9	1.0
.1	.090	.044	.024	.019	.012	.010	.009	.007	.006	.006	.005
.2	.159	.134	.099	.093	.049	.039	.033	.029	.025	.022	.020
.3	.236	.221	.174	.132	.104	.095	.072	.092	.055	.049	.044
.4	.311	.301	.263	.214	.174	.145	.124	.108	.095	.085	.077
.5	.393	.376	.349	.300	.253	.215	.189	.163	.144	.130	.118
.6	.451	.449	.425	.385	.336	.291	.255	.225	.201	.191	.165
.7	.516	.511	.496	.463	.417	.370	.329	.293	.263	.238	.217
.8	.576	.573	.560	.535	.494	.447	.403	.363	.329	.299	.274
.9	.632	.629	.619	.599	.565	.521	.479	.433	.395	.362	.333
1.0	.693	.690	.672	.657	.629	.590	.546	.503	.462	.426	.393
1.1	.729	.727	.720	.709	.696	.652	.612	.569	.527	.489	.454
1.2	.770	.769	.763	.753	.736	.709	.671	.631	.599	.550	.513
1.3	.906	.905	.901	.793	.779	.757	.725	.697	.647	.609	.570
1.4	.839	.937	.934	.928	.917	.799	.772	.739	.701	.662	.625
1.5	.866	.866	.893	.859	.949	.835	.813	.783	.749	.712	.675
1.6	.990	.890	.997	.993	.977	.969	.949	.923	.792	.757	.722
1.7	.911	.910	.909	.905	.900	.891	.877	.956	.929	.799	.764
1.8	.928	.929	.929	.924	.920	.913	.902	.885	.961	.933	.902
1.9	.943	.942	.941	.939	.936	.931	.922	.909	.899	.894	.939
2.0	.954	.954	.953	.952	.949	.945	.939	.929	.912	.990	.965
2.1	.964	.064	.963	.992	.960	.957	.952	.944	.930	.912	.890
2.2	.972	.972	.972	.971	.969	.967	.963	.957	.946	.931	.911
2.3	.979	.978	.979	.877	.976	.974	.972	.967	.958	.946	.929
2.4	.994	.994	.993	.993	.992	.981	.979	.875	.969	.959	.944
2.5	.999	.999	.997	.987	.986	.985	.994	.981	.976	.969	.958
2.6	.991	.991	.990	.990	.990	.989	.899	.896	.982	.876	.966
2.7	.993	.993	.993	.993	.992	.992	.991	.990	.997	.992	.974
2.8	.995	.995	.995	.895	.994	.994	.993	.992	.990	.989	.980
2.9	.996	.996	.996	.999	.996	.999	.895	.894	.993	.990	.985
3.0	.997	.997	.997	.997	.997	.997	.997	.999	.995	.993	.999
3.1	.998	.999	.999	.998	.999	.999	.998	.997	.986	.995	.992
3.2	.999	.999	.999	.998	.999	.998	.999	.999	.997	.996	.994
3.3	.999	.999	.999	.899	.999	.899	.999	.999	.999	.997	.996
3.4	.999	.999	.999	.999	.999	.999	.999	.999	.999	.998	.997
3.5	1.000	1.000	1.000	1.000	.999	.999	.999	.999	.999	.999	.999
3.6	1.000	1.000	1.000	1.000	1.000	1.000	1.000	1.000	1.000	.998	.999
3.7	1.000	1.000	1.000	1.000	1.000	1.000	1.000	1.000	1.000	.999	.998
3.8	1.000	1.000	1.000	1.000	1.000	1.000	1.000	1.000	1.000	1.000	.999
3.9	1.000	1.000	1.000	1.000	1.000	1.000	1.000	1.000	1.000	1.000	1.000
4.0	1.000	1.000	1.000	1.000	1.000	1.000	1.000	1.000	1.000	1.000	1.000

Values from the Computing Laboratory, Ballistic Research Laboratories.

of force-in-being or potential force by making one or more components of these forces the target of the missiles system and by inflicting damage on these components. The target of a particular missile therefore may or may not be part of an attacking force. An included purpose is to inflict attrition on the enemy's force and thus weaken his capability for attack.

In some cases it is important to inflict attrition-type damage which will add to the burden of repairs at the enemy's base. This may be true in a long air campaign in which enemy effort is limited by aircraft availability rather than by weapon or target availability. In general, this is not important for naval applications where short campaigns are more common. However, it might enter to some extent in major amphibious operations.

These various types of damage are collectively called "kills" even though the word literally describes only part of the cases. To differentiate between various types, a specialized nomenclature has developed. The most commonly used categories are:

Aircraft Kills

KK-damage is immediate catastrophic disintegration of the target, thus completely eliminating this component of the enemy's force.

K-damage is damage which completely defeats the target "immediately" (usually interpreted as within 10 seconds), thus eliminating this component of the enemy's force immediately as a target.

A-damage is damage which completely defeats the target within 5 minutes, thus allowing this component 5 minutes of potential use before it is eliminated from the enemy's force.

B-damage is damage which completely defeats the target within 2 hours, thus allowing this component 2 hours of potential use before it is eliminated from the enemy's force.

C-damage is damage which prevents the

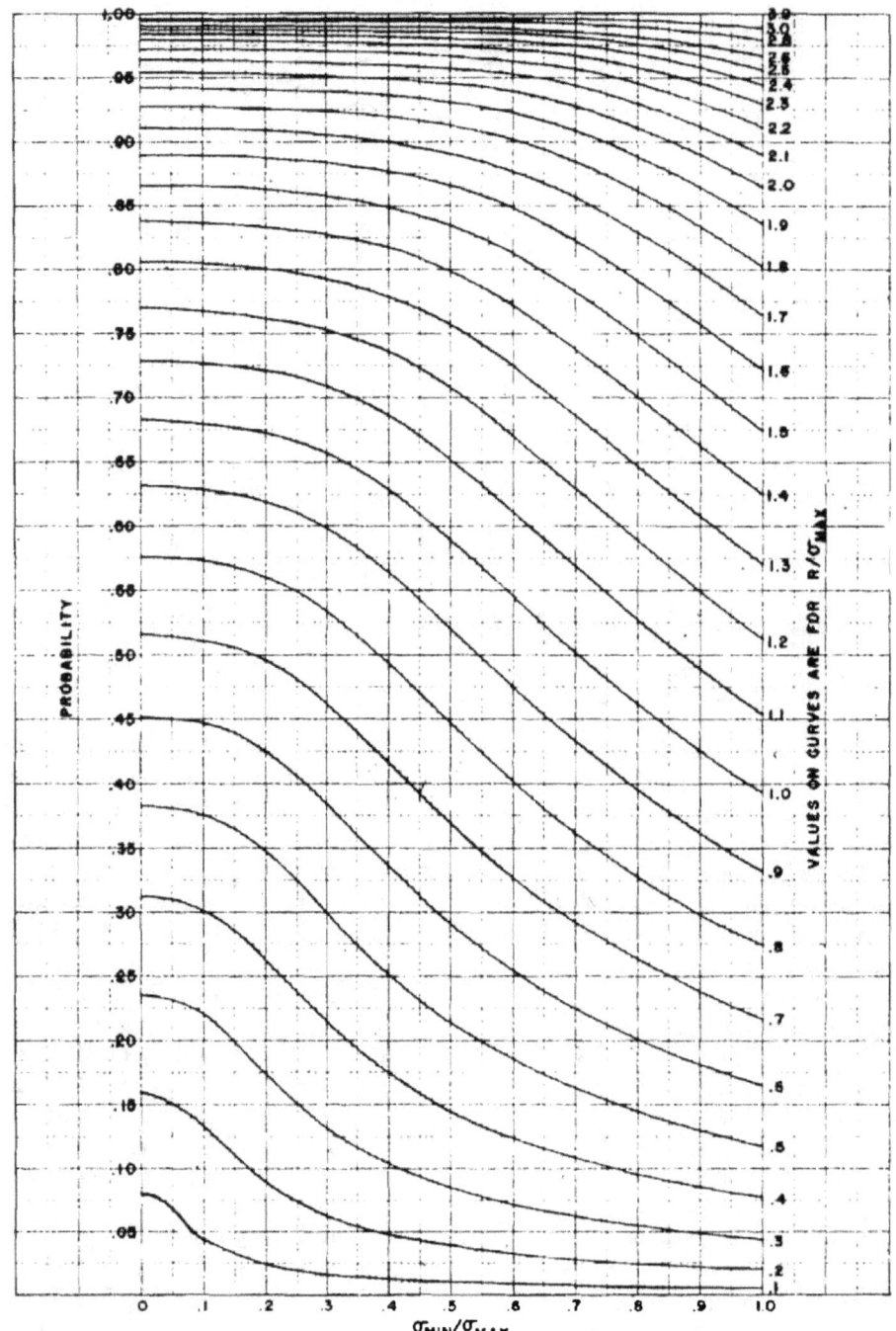

Figure 5-9. Cumulative Bivariate Normal Distribution (Over Circles of Radius R Centered at the Mean)

target from carrying out its primary mission. Thus, this component of the enemy's force may still be employed in a secondary use or may be re-employed in its primary mission after a period of time. It may also require repairs before re-employment in its primary mission.

D-damage occurs when more than a given number of man hours is required for repair of damage inflicted to the target before it can be re-employed by the enemy.

E-damage is damage which prevents the target from being available for at least the next scheduled mission.

It should be noted that at least C-damage is always desired.

Kills On Armored Vehicles

The classification of kills on armored vehicles is made on a different basis than for aircraft. For the details of standardization, reference may be made to the reports of the Fourth Tripartite Conference on Armor. Briefly indicated, the following kills are used:

K-damage Complete destruction of vehicle.
F-damage Loss of fire power.
M-damage Loss of mobility.

Kills Against Personnel

The criteria for incapacitation of normally clothed but otherwise unprotected troops are considerably more complicated than for machinery and, as may be expected, suffer from a lack of experimental determination. The most recent fundamental work is contained in reference 5-hh. These criteria depend on the duties of the troops involved and the time required for incapacitation. Older criteria such as the German one; namely, 58 ft-lb of energy in a shell or fragment being lethal and criteria based on penetration depth in the body have been superseded. The probability of incapacitation, P_{bk}, is now expressed as

$$P_{bk} = 1 - e^{-a(mV^{1/2} - b)^n} \qquad (5-1.23)$$

where m and V are the mass and velocity of the striking material and the constants a, b, and n are tabulated for the personnel involved according to their duties and time-to-incapacitate.

To approach a realistic figure of target vulnerability for warhead evaluation purposes, the desired level of damage must be decided upon. This decision is influenced to a large degree by the missile to be performed by the target. For example, defense against a Kamikaze attack requires KK-damage to the attacking aircraft, while A-damage to a bomber would be considered sufficient if the bomber were over five minutes from the bomb release point. A bridge being used by an enemy force may require A-damage while B-damage would be sufficient to an industrial installation, say an oil refinery.

5-2. FUNDAMENTAL CONCEPTS

5-2.1. The General Concept As shown by equation 5-1.2 the conditional kill probability which is of concern to an evaluator of warhead effectiveness is

$$P_k = f[\phi(G), \psi(F), V_m, V_t, \theta, h, l(m), V(T)] \qquad (5-2.1)$$

It remains to be shown how each of the above factors affect P_k.

Each warhead type has a unique method of inflicting damage on a target. The blast warhead depends upon a shock wave with high impulse and overpressure. The fragment warhead emits a front of small, high-velocity projectiles; the rod warhead, a front of steel bars. The cluster warhead emits expanding rings of submissiles that damage like other types of warheads, usually blast. The incendiary uses heat. The shaped charge warhead is essentially a gun which shoots out an extremely high-velocity molten metal mass. The chemical warhead uses toxic chemical substances and the biological depends on live disease organisms or their toxic products.

In every case the evaluator wants to know for a randomly fired missile; (1) the probability that the damaging agent reaches the target, (2) what are the characteristics of the damaging agent when it reaches the target, and (3) how effective are those characteristics in inflicting the desired damage.

When missiles are fired at a target, their warheads will burst at various positions around the target and some may even hit the target. A picture showing six such bursts distributed simultaneously around an airplane target is given in Figure 5-11.

Assuming perfect functioning of the warhead, the reason these bursts occur away from aim-point is that there are guidance and fuzing errors in the missile system. These errors were discussed in Sections 5-1.4 and 5-1.5. Once the evaluator knows the distributions of these errors, he can determine the probability that the warhead will burst at any particular position.

For each position there is a definite probability of effecting a kill. This probability depends upon the characteristics of the damaging agent when it reaches the target and the effectiveness of these characteristics against the target.

P_k can now be expressed in terms of the probabilities of bursts at N various positions; i.e., centroids of units of volume equally likely to contain a burst, and the probabilities of kill when warheads burst at these positions. Mathematically this is

$$P_k = \sum_{i=1}^{n} p_i \cdot p_{ki} \quad (5-2.2)$$

where:

p_i = probability of bursting at the i'th position (centroid of unit of volume equally likely to contain a burst) = $\frac{1}{n}$

p_{ki} = probability of kill given a burst at the i'th position.

Now, pi is dependent only on guidance and fuzing errors. Thus,

$$p_i = f\left[\phi(G), \psi(F)\right] = \frac{1}{n} \quad (5-2.3)$$

Also, p_{ki} is dependent only on the remaining factors given in equation 5-2.1 and can be expressed by

$$p_{ki} = f\left[V_m, V_t, \theta, b, l(m), V(T)\right] \quad (5-2.4)$$

Thus, the effects of $\phi(G)$ and $\psi(F)$ on P_k have been shown and separated from equation 5-2.1 by the geometry of the centroids. It now remains to show how the factors in equation 5-2.4 affect kill-probability. These effects are different for each warhead type.

5-2.2. Blast Warhead   With a blast warhead, the evaluator is concerned with what overpressure and impulse are applied on the target. The effective overpressure and impulse may be determined by knowing the initial values and the distance the shock wave must travel to reach the target.

The shock wave moves at supersonic velocity. Even so, the time of travel of this shock wave is appreciable. This time is used along with that due to the target velocity (in the case of an airborne target) while the target

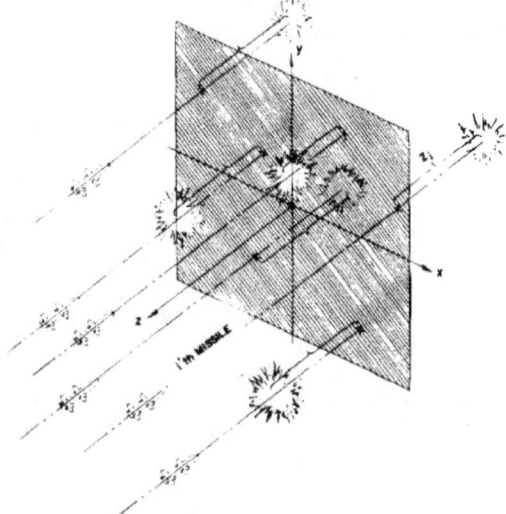

Figure 5-10. Random Fuzing Errors Combined with Random Guidance Errors

moves to intercept the blast wave. The evaluator needs to calculate the distance, d, between the burst-point and the target at time of intercept of the blast wave as illustrated in Figure 5-12 for an aerial target. Once this distance is determined he may then modify the original lethality in terms of overpressure and impulse to effective lethality in terms of effective overpressure and effective impulse.

The effect of these modified values is determined by the target vulnerability. The latter varies with the target and its parts. The tail is usually most vulnerable. A blast-kill envelope is usually made for each target and charge weight.

5-2.3. Fragment Warhead With a fragment warhead, the evaluator is concerned with the number, size and velocity of the fragments that hit the target. These values vary according to the distance, d, between the burst point and the position of the target at the time the fragments intercept it. This is shown in Figure 5-13.

The lethality of a fragment depends considerably on its striking velocity, V_s, at the target and, therefore, a knowledge of V_s is important. It is related to the initial dynamic ejection velocity of the fragment, V_d, the rate of slow-down, the distance, y, and the target velocity, V_t. (See Figure 5-14.)

For the treatment below the following usually adequate approximations are made, most of which are based on the fact that the flight-time of the fragment is generally less than one second:

Assumptions
1. The air density, ρ_a, is a constant.
2. The path of the target is a straight line.
3. Gravitational effects are neglected.
4. The drag coefficient, C_D, of the fragment is constant.

A small error will exist due to assumption four if the fragment passes through the sonic speed region during the flight.

Before discussing the geometry of the situation, the velocity history of the fragment with respect to the air should be given. From the basic force equation

$$M \frac{dV_x}{dt} = - C_D \rho_a A V_x^2 ; \qquad (5-2.5)$$

setting $\frac{dV_x}{dt} = V_x \frac{dV_x}{dx}$,

one obtains

$$V_{xs} = V_d e^{\frac{-C_D \rho_a A}{M} x_s} \qquad (5-2.6)$$

where C_D, A, and M refer to the fragment ballistic drag coefficient*, presented area and mass respectively. Equation 5-2.6 describes the downrange velocity of fragments launched with dynamic velocity V_d. If consistent units are used, the exponent $C_D \rho_a A x_s / M$ is dimensionless.

*This is one-half the drag coefficient commonly used in aeronautical work.

In order to define a convenient coordinate system, one may use a plane which contains the path of the target and the burst point of the warhead. The path of the fragment will also lie in this plane. As soon, therefore, as the various velocities which contribute to V_d (i.e., those due to the explosion plus warhead motion) are resolved, the problem is reduced to two dimensions.

Consider first, V_f, the initial velocity of the fragment relative to the missile, provided by the detonation. This velocity is composed of two components, V_R radially and V_g longitudinally. The sum of V_g and the missile velocity, V_m, then give the fragment velocity along the missile axis, which need not lie in the plane described. Introducing θ_s, the angle at which the fragment is ejected (measured from the missile axis), the steps are:

$$V_R = V_f \sin \theta_s \qquad (5-2.7)$$

$$V_g = V_f \cos \theta_s \qquad (5-2.8)$$

$$V_d = \sqrt{V_R^2 + (V_g + V_m)^2} \quad \text{(Vector sum)} \qquad (5-2.9)$$

The velocity V_d must lie in the above-described plane. This fact determines θ_s. The problem now appears as shown in Figure

5-14, where the velocity of the fragment (in free air) during its approach to the target is coplanar with the target path and given by equation 5-2.6.

The initial distance w and angle β are known, as is the target velocity V_t. The final desired value is the resultant striking velocity, V_s, which is found by using the Law of Sines:

$$V_s = \sqrt{(V_{xs} \cos \theta_d - V_t)^2 + (V_{xs} \sin \theta_d)^2} \quad (5\text{-}2.10)$$

In addition, for intercept, the flight time of the target and fragment must be equal. Thus

$$\frac{x_s}{\overline{V}} = \frac{u + Z}{V_t} \quad (5\text{-}2.11)$$

where $\overline{V}$ is the average velocity of the fragment in free air. This is obtained by taking the time integral of equation (5-2.6) over the range x_s and dividing by $x_s/\overline{V}$. Thus

$$\overline{V} = \frac{V_d \alpha x_s}{e^{\alpha x_s} - 1} \quad \text{where} \quad \alpha = \frac{C_D A \rho_a}{M} \quad (5\text{-}2.12)$$

The geometry of the attack provides the following relations (Figure 5-14):

$$y^2 + z^2 = x_s^2 \quad (5\text{-}2.13)$$

$$y^2 + u^2 = w^2 \quad (5\text{-}2.14)$$

$$z = x_s \cos \theta_d \quad (5\text{-}2.15)$$

$$u = w \cos \beta \quad (5\text{-}2.16)$$

The situation may be summarized as follows: The known parameters are V_d, V_t, y and β. There are eight unknowns; namely, V_{xs}, V_s, $\overline{V}$, x_s, θ_d, u, w, and z. Corresponding to these are eight equations; namely, equations (5-2.6) and (5-2.10) through (5-2.16) which together enable a solution for V_s to be obtained. The four geometrical relations reduce to

$$w \sin \beta = x_s \sin \theta_d \quad (5\text{-}2.17)$$

and equation (5-2.11) may be rewritten, using the geometrical relations, as

$$\frac{x_s}{\overline{V}} = \frac{w \cos \beta + \sqrt{x_s^2 - w^2 \sin^2 \beta}}{V_t} \quad (5\text{-}2.18)$$

The combination of 5-2.18 and 5-2.12 gives x_s. The value of x_s obtained can be used in equation (5-2.6) to solve for V_{xs} and in equation (5-2.17) to get θ_d. The values of x_s and θ_d are then substituted into equation 5-2.10 to obtain the relative striking velocity V_s.

For a more accurate approximation of the velocity vector relationships, see Reference 5-gg, which provides an averaging (approximation) for different longitude angles all around the warhead, taking the axis of the warhead in the V_m direction as the polar axis in terms of the terrestrial sphere.

5-2.4. Rod Warhead With rod warheads, the evaluator is concerned with how many bars of what size hit the target at what effective velocity. These values vary according to the distance, d, between the burst point and the position of the target at the time the bars intercept it as shown in Figure 5-15.

The method is the same as that used for fragment warheads. The main differences are that, from the rod warhead, the number of projectiles is smaller, they are moving at a lower velocity, and their rate of slow-down is somewhat different than for usual fragments (depending on the fragment mass and shape). Because of these lower velocities, the velocity-vector relationships are of greater significance. These velocity-vector relationships are dependent upon V_m, V_t, θ, and the design characteristics of the warhead.

5-2.5. Cluster Warhead In evaluating a cluster warhead with submissiles equipped with an "all-ways fuze", one is concerned with whether or not one or more of the submissiles hit the target. If a submissile does hit and fire, it is assumed that there is a kill. However, it is to be noted that submissile fuzes have a lower reliability, in general, than warhead fuzes. If no submissiles hit, then there is no kill except in the case of submissiles using time or proximity fuzes, in which case a hit of the submissile on the target is not necessary for a kill.

The former situation is illustrated in Figure 5-16. The distance, d, between the position of missile burst and the position of the target at the time of the submissile pattern intercept is much larger than for other types of warheads. As with the fragment and rod warheads, the time of travel over distance, d, is important. Knowing this time, the target is moved from its position at time of missile burst (shown in outline) to its position at time of submissile intercept (shown in solid).

Having obtained (by the technique described in Section 5-2.3) the modified position of the target along with information on the submissile pattern, the evaluator may then determine the hit probability of at least one submissile. This in turn is equal to P_{k1} for perfect submissile fuze operation.

In the case of time or proximity fuzed submissiles, the evaluator must determine the distance from each submissile burst to the target and then determine the effective lethality in terms of effective overpressure and effective impulse. The results are then determined in a manner similar to that used for blast warheads.

5-2.6. Shaped Charge Warhead The jet of metal from a shaped change must, of course, strike the target directly. Against ground targets this generally means that the warhead detonates on contact with the target surface. Against air targets, the detonation may occur at distances so large that standoffs are of the order of a hundred feet. In this case, the axis of the warhead must be aligned critically to intersect the target. Since the jet of metal travels at speeds of Mach 10 to 20, the target velocity V_t, missile velocity V_m, and distance traversed by the jet are relatively unimportant (provided the latter is in the range of effectiveness of the weapon) for computing whether the jet intercepts the target. (For such targets and long standoffs, the "jet" becomes more like a strung-out and wide spray of rapidly moving metal particles.)

There are corresponding differences in evaluating damage, once a hit has occured. The effect of a large shaped charge jet on aircraft-type structure, see Figure 1-31, is so violent that a hit on the fuselage or wings, with the possible exception of the extreme tips, is generally considered to produce A, if not K, damage. On the other hand, damage to tanks and fortifications must be evaluated much more carefully. There are two main judgements to be made. First, did penetration of the protective shell occur. If not, then in most incidences only minor damage has been suffered. If penetration has occurred, then it must be determined what effect the spall and hot gases have had on the target occupants and the soft components of the target interior. Damage to these components is usually done by small fast moving flakes of wall material which produce holes in hydraulic lines, communications equipment and personnel. A close examination by the evaluator is generally required to determine the extent and class of kill.

5-3. APPROXIMATE EVALUATIONS

5-3.1. General Discussion This subchapter presents simplified evaluations for certain types of warheads against aerial targets. In a step by step process the reader may pick each important parameter and proceed to an approximate value of kill probability. The data which has been used in compiling this graphical material was gathered from a large number of references as indicated in each subchapter section. Reference 5-ee is typical of these. There are, of course, a number of ways in which the basic parameters may be combined into variables and plotted; the particular steps shown are not unique.

Because sufficient data is unavailable in the present state of the art for other types of warheads, only external blast, internal blast, and fragment warhead approximate evaluations are included. These are given in Figures 5-18 through 5-30. To use these graphs, one simply starts with the warhead weight and follows through the set until he arrives at an approximate evaluation interval.

It should be desirable to have sufficient data to follow this approach for all types of warheads against both land and aerial targets. It is hoped that future evaluation work will supply these data.

5-3.2. External Blast Warheads The approximate evaluations of external blast warheads against some aerial targets are found by using Figures 5-18 to 5-22. These are based on the assumptions of perfect fuzing and a circular normal distribution of guidance error as discussed in Section 5-1.4. The basic information was obtained from References 5-c through 5-l and Reference 5-ff.

An equivalent method to using Figures 5-18 to 5-22 may be found on page 25, Figures 8 and 9 of reference 5-gg. Use of Figures 5-18 and 5-19 is essentially the same as use of the empirical relationship $\frac{\sigma_G}{W^n}$ where n varies between .4 and .5 and where the former value corresponds to the large bomber. Dr. P. Whitman of APL-JHU found that $n = .5$ holds for point targets.

5-3.3. Internal Blast Warheads Approximate evaluations of internal blast warheads against some aerial targets are found by using Figures 5-23 through 5-26. These are based on information from References 5-l through 5-v. Perfect fuzing and a circular normal distribution of guidance error are assumed.

5-3.4. Fragment Warheads Approximate evaluations of fragment warheads against some aerial targets may be found by using Figures 5-27 through 5-30. These are based on information obtained from references 5-w through 5-gg. Perfect fuzing and a circular normal distribution of guidance error are assumed.

5-4. EVALUATION METHODS

5-4.1. Analytical Method Basically, the evaluation problem is to determine as well as possible the adequacy of a developed warhead relative to the original requirements to which it was developed. This is done by determining the conditional kill probability, P_k; i.e., the probability of inflicting specified damage (kill) upon a given target provided that the target is detected, the missile system functions, the warhead is delivered to the target, and the fuzing system functions. The relationship of this conditional kill probability to the overall kill probability of the missile system is discussed in Section 5-1.2 and set down in equation 5-1.2.

Besides the techniques of full-scale or approximate evaluation, it is possible to construct mathematical methods for determining the effectiveness of warheads. Since the physical phenomena on which warhead actions depend contain many random variables, it follows that these mathematical methods are based on the theory of stochastic processes and are especially designed for handling statistical problems; some in fact are aided by graphical or mechanical assists such as random number tables, card files, dice, and so forth.

Some of the fundamentals of these techniques are described below in general terms. However, there is sufficient variability between evaluation problems that the discussion should be considered as a guide only and the particular format of evaluation employed is usually tailored to the problem.

The component parts of the conditional kill probability are discussed in general in Section 5-1.3. As shown in that section, the conditional kill probability is

$$P_k = f[\phi(G), \psi(F), V_m, V_t, \theta, b, l(m), V(T)] \quad (5\text{-}4.1)$$

where:

$\phi(G)$ = frequency distribution of the guidance error

$\psi(F)$ = frequency distribution of the fuzing error

V_m = velocity of the missile

V_t = velocity of the target

θ = angle between missile and target trajectories

b = altitude of engagement

$l(m)$ = missile lethality

$V(T)$ = target vulnerability.

The problem now becomes one of analyzing the above factors and relating them to P_k.

The conditional kill probability can also be written as the product of other probabilities; i.e.,

$$P_k = P_b \cdot P_c \cdot P_z \cdot P_d. \quad (5\text{-}4.2)$$

where:
- P_b = probability of a hit on some specific area normal to the missile trajectory
- P_c = probability of kill provided that area is hit
- P_z = probability of fuzing within a distance, z, of the aim point
- P_d = probability of a kill provided the warhead fuzes within a distance z.

Where it is desired to obtain the conditional kill probability for a salvo or multiple bursts of any type, it is of course necessary to sum the probabilities over the various burst volumes. There are, in fact, many additional complexities which may arise in a particular situation. A typical one would be the joint kill probability which arises for P_c in the case of a multiply-vulnerable target where the kill due to a hit on a pilot may depend on whether there has also been a hit on the copilot. Many similar situations may be cited.

The probability P_{kn} of killing a target with n bursts of individual kill probability P_k is, in general, $P_{kn} \leq 1 - (1-P_k)^n$. The upper limit is approached as the complexities have less effect.

The probability of a hit on a specified area is covered in Section 5-1.4., Distribution of Guidance Error. It is usually assumed that guidance error is distributed normally and independently in the x and y directions. If this assumption is made and the area is rectangular in shape, then P_b is found by equation 5-1.12; i.e.,

$$P_b = P_a \cdot P_b$$

where:
- P_a = probability that the x is between $\pm a$
- P_b = probability that the y is between $\pm b$

and by using Table 5-1 to find P_a and P_b. Thus, for a rectangular presented area, the conditional kill probability is

$$P_k = P_a \cdot P_b \cdot P_c \cdot P_z \cdot P_d. \quad (5\text{-}4.3)$$

If the area is circular of radius, m, from the aim point then

$$P_b = P_\Delta \quad (5\text{-}4.4)$$

where:
- P_Δ = probability of hitting within a circle of radius m.

If it is assumed that $\sigma_x = \sigma_y = \sigma_G$ (circular normal distribution), then P_Δ is given by equations 5-1.16 and 5-1.18 or P_Δ can be found by using Figure 5-8 or Table 5-2.

Depending on the type of weapon delivery system, guidance, and other factors, the standard deviations in the x and y directions, σ_x and σ_y may or may not be equal. Where these are even approximately equal, one can use the "circular" deviation

$$\sigma_G = \sqrt{\sigma_x^2 + \sigma_y^2}$$

It is assumed that $\sigma_x \neq \sigma_y$ (elliptical normal distribution) and the difference is large, then P_Δ can be found by using either Table 5-3 or Figure 5-9.

Thus for a circular presented area, the conditional kill probability is

$$P_k = P_\Delta \cdot P_c \cdot P_z \cdot P_d. \quad (5\text{-}4.5)$$

The fuzing probability for a particular fuze type is discussed in Section 5-1.5, is given by equation 5-1.22 and is found by using Table 5-1.

This now leaves two of the factors of equation 5-4.2 unknown, P_c and P_d. The probability of kill provided the area is hit is

$$P_c = f_1 [W, V_m, V_t, \theta, b, l(m), V(T)]. \quad (5\text{-}4.6)$$

and the probability of kill provided the warhead fuzes within a distance, z, of the plane through the aim point and perpendicular to the missile path, is

$$P_d = f_2 [W, V_m, V_t, \theta, h, l(m), V(T)], \quad (5\text{-}4.7)$$

where:

f1 and f2 are different functions of the same variables.

The relationships of these variables to P_c and P_d are not usually amenable to mathematical expression. As a result, methods have been developed to estimate P_c and P_d, such as "Simulated", "Graphical", and "Overlay". These methods are discussed in the following sections.

5-4.2. Simulated Method Frequently it is impossible to obtain the actual targets against which the warhead is to be evaluated. Often, even if the actual targets may be obtained, it is undesirable or too expensive to evaluate on the basis of actual test firings. For these reasons the "Simulated Method" has been devised to aid in estimating the evaluation.

Basically, the simulated method amounts to building a model of the target, using a random process to determine approximately a hundred (in some cases many more) firings, computing the kill probability for each of these firings, and then using these data to arrive at conditional kill probabilities. For low values of P_k a relatively large number of "experiments" are required to obtain accurate values, since the number required varies inversely with the value of P_k.

The random process most frequently used to determine the path of the missile, fragment, rod, shaped charge, or submissile is based on the "Monte Carlo" Method or the "Lotto" Method. These methods are discussed in Section 5-4.3.

The devices that have been used to project the path to and/or through the target are many. Some of these are rods, light rays, and gamma rays. When rods are used, they are placed in the position of the path and touch the model. If they pass through a vulnerable component without first hitting a shielding component, then a hit or kill is recorded dependent upon the lethality of the object presumably following the path and the vulnerability of the component.

When the light rays or gamma rays are used, the evaluator is mainly interested in determining the number of hits on the presented area; i.e., whether the rays touch the target. He finds the probability of kill for the area hit. Using the frequency of hits, he estimates the probability of a hit there. The conditional kill probability is the average product of these two probabilities for many trials.

5-4.3. Monte Carlo and Lotto Methods The Monte Carlo technique is a mathematical tool which was developed during the 1940's for performing analyses of physical phenomena which obey the laws of random processes. Principal among these are problems in neutron diffusion and other atomic effects. But any phenomena which occur according to a known distribution function may be examined by a similar scheme.

In the Monte Carlo technique, one does not examine each individual event but instead one generates mathematically a "typical" sample of an assembly of events (such as for example the location of bursts in a salvo). These samples

Figure 5-11. Random Warhead Bursts Around Target

are produced by using a table of random numbers or some mechanical device whose output is distributed in the proper way. To suit the problem at hand, proper corrections must be made for mean value, standard deviation and weighting, if necessary.

Since a statistical sample serves to represent the entire array of events, it is essential to know the fidelity of the sample. For binomial distributions the standard error of the estimated probability which may occur is usually expressed as $\sigma_p \simeq \sqrt{p(1-p)/n}$ where p is the estimated probability of the event under study and n is the number in the statistical sample. For normal distributions this error is expressed by

$$\sigma_p \simeq \sqrt{p(1-p)/n-1} \ .$$

As Monte Carlo type data is applied to warhead evaluation, the desired accuracy is determined by use of the data. For example, in determining the relative merits of two very different types of attack, low accuracy (and a correspondingly small sample) may suffice. On the other hand, for comparison of similar weapons or the optimization of parameters, much larger samples are required—especially when the kill probability per attack is small. The size of the statistical sample together with the particular problem determines the confidence which can be placed in the result.

The Lotto method is an extension of the Monte Carlo principle that is especially adapted to the study of random phenomena as they occur in calculations of weapon effectiveness against multiply-vulnerable targets. Using the example previously referred to, a random device or list is used to produce a statistic sample of miss distances. Similarly geometric position of the sample miss may be obtained. Then by examining a model of the target, the presented area of each vulnerable component may be determined, e.g., for a fragmenting warhead. Using the range and area, a table may be entered to determine the kill probability for any vulnerable component. Finally another random table may be entered to determine, from a yes-no decision, if the component was actually killed by a fragment from the burst. This procedure can be carried out for a number of bursts and a number of vulnerable components, keeping score of that fraction of the total bursts which killed sufficient components to produce a kill of the target or targets.

Included in the Bibliography, Subchapter 5-6, is a list of references to the Monte Carlo and Lotto methods and their application to warhead testing.

5-4.4. Overlay Method The overlay method of evaluation involves basically reducing the target to two dimensions on a plane, laying over this plane a grid with points, and evaluating the damage at each point.

The original picture on the plane may be obtained by photographing a target or a model of the target. If the target should be personnel

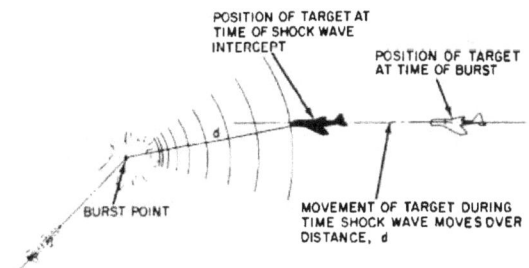

Figure 5-12. Critical Distance, d, for Evaluation of Blast Warhead

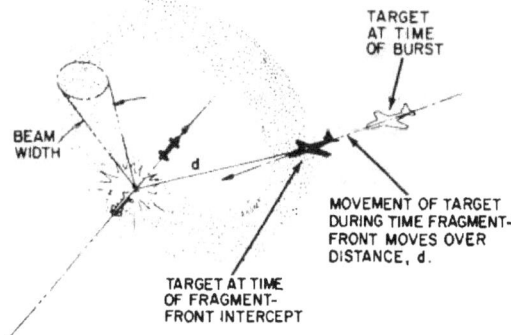

Figure 5-13. Critical Distance, d, for Evaluation of Fragment Warhead

of known distribution, circles may be randomly located by the Monte Carlo method to represent individual persons.

The points on the overlay grid are determined from a known or predicted distribution and, sometimes, with the aid of the Monte Carlo method.

5-4.5. Graphical Method The Graphical Method is based on data gathered from tests and known or predictable distributions. It amounts to plotting on a set of graphs the points that have been found, fitting curves to these points, and then using these curves to effect the evaluation.

Some of the curves may be theoretical and used in combination with test curves. This is basically the method used in the approximate evaluation graphs of Subchapter 5-3.

5-4.6. Geometrical Model Method The Geometrical Model Method is one where the vulnerable components of the target are defined by simplified geometrical shapes and located in either a Cartesian plane or Euclidian (3-dimensional) space. These geometrical equations are then fed into a computer where a random method is used along with mathematically defined distributions for the warheads or their projectiles.

This method lends itself readily when computer facilities are available and adequate information is available to define geometrically the distributions and shapes.

5-5. REFERENCES

5-a "Armament", Harold Goldberg, published in "Principles of Guided Missile Design", edited by Grayson Merrill, Van Nostrand, Princeton, 1956.

5-b "Standard Mathematical Tables", Chemical Rubber Publishing Co., Cleveland, Ohio, 1954.

5-c "Vulnerability of A-35 Aircraft to External Blast", E. Gilinson, BRL Tech. Note 277, August, 1950.

5-d "Damage to Aircraft by External Blast", W. E. Baker, BRL Report 741, October, 1950.

5-e "Damage to B-17 and B-29 Aircraft by External Blast", W. E. Baker and O. T. Johnson, BRL Memo. Report 561, September, 1951.

5-f "Vulnerability of B-29 Aircraft to 120 M.M. Air-Burst Shell", Johns Hopkins University, Project THOR, Tech. Report 3, February, 1950.

5-g "Report on Tests of the Effect of Blast from Bare and Cased Charges on Aircraft", James N. Sarmousakis, BRL Memo. Report 436, July, 1946.

5-h "The Effect of Blast on Aircraft Reciprocating Engines", Arthur Stein and Harry Kostiak, BRL Memo. Report 467, August, 1947.

5-i "The Effect of Blast on Aircraft Fuel Tanks", Frances M. Hill, BRL Memo. Report

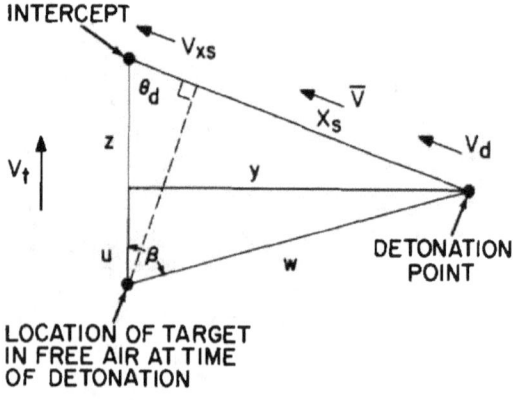

Figure 5-14. Geometry for Fragment Striking Velocity

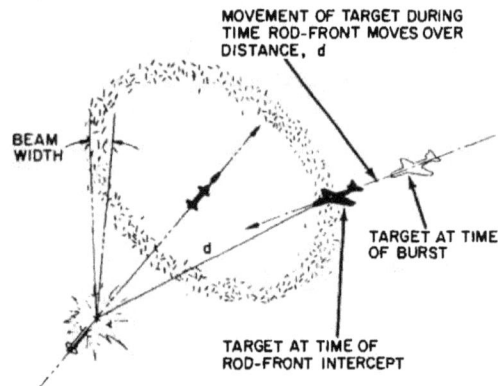

Figure 5-15. Critical Distance, d, for Evaluation of Rod Warhead

509, April, 1950.

5-j "The Effect of Altitude on the Peak Pressure in Normally Reflected Air Blast Waves", A. Hoffman, BRL Tech. Note 787, March, 1953.

5-k "Damage to B-47 Aircraft by External Blast", O. T. Johnson and R. T. Shanahan, BRL Memo. Report 736, October, 1953.

5-l "The Effect of Atmospheric Pressure and Temperature on Air Shock", J. Dewey and J. Sperrazza, BRL Report 721.

5-m "The Effect of Blast on Aircraft", Joseph Sperrazza and James N. Sarmousakis, BRL Report 645, August, 1947.

5-n "Vulnerability of the Type 39 Aircraft to Internal Blast", James J. Dailey and Sarkis E. Giragosian, BRL Memo. Report 980, March, 1956.

5-o "Vulnerability of B-29 Aircraft to Internal Blast", Joseph Sperrazza, BRL Memo. Report 490, June, 1949.

5-p "Vulnerability of B-29 Aircraft to Eight-Pound Bare Charges Fired Externally", P. N. French and D. W. Mowrer, BRL Memo. Report 942, October, 1955.

5-q "The Damage Effect of Small TNT Bare Charges Placed Inside Standard and Modified B-17 Fuselages", J. N. Sarmousakis, BRL Memo. Report 442, August, 1946.

Figure 5-17. Critical Angle, θ, for Evaluation of Shaped Charge Warhead

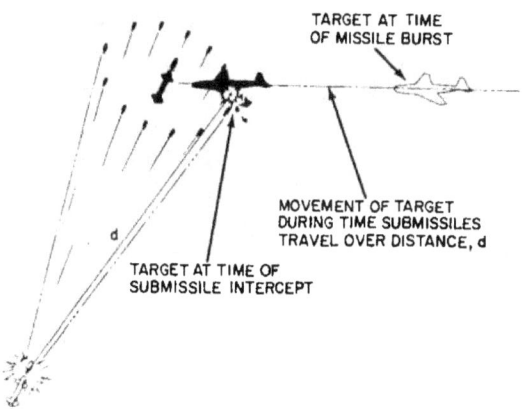

Figure 5-16. Critical Distance, d, for Evaluation of Cluster Warhead

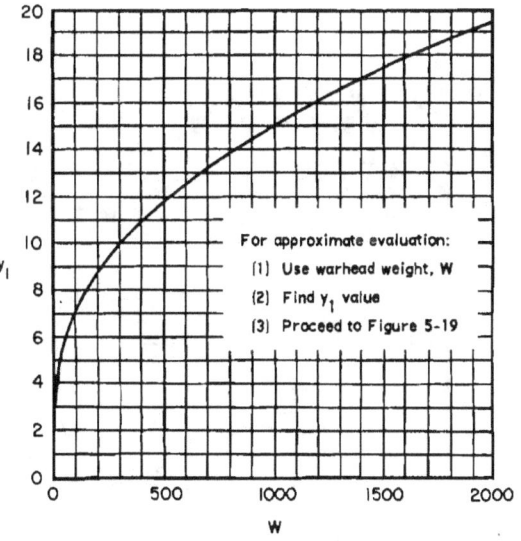

Figure 5-18. External Blast Warhead Evaluation - Warhead Weight Variable

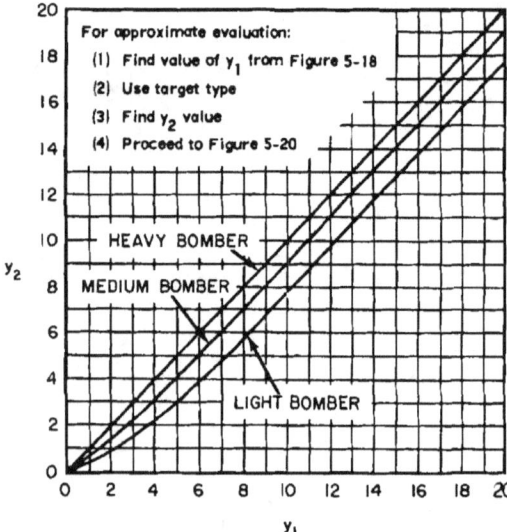

*Figure 5-19. External Blast Warhead Evaluation-
Target Type Variable*

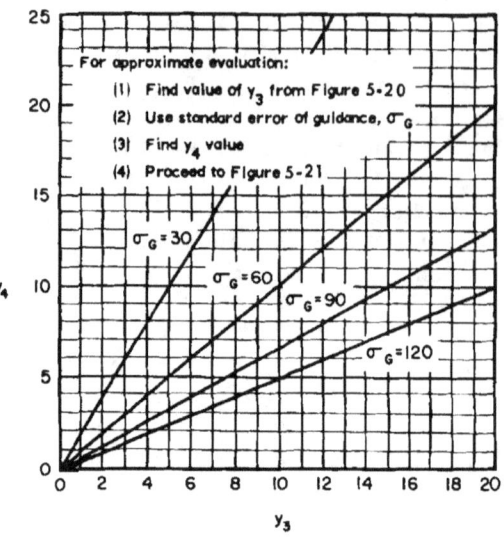

*Figure 5-21. External Blast Warhead Evaluation-
Standard Error of Guidance Variable*

*Figure 5-20. External Blast Warhead Evaluation-
Engagement Altitude Variable*

*Figure 5-22. External Blast Warhead Evaluation-
Kill Probability Intervals*

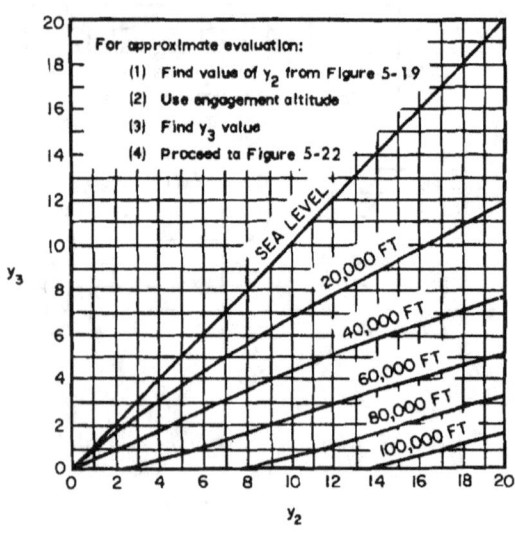

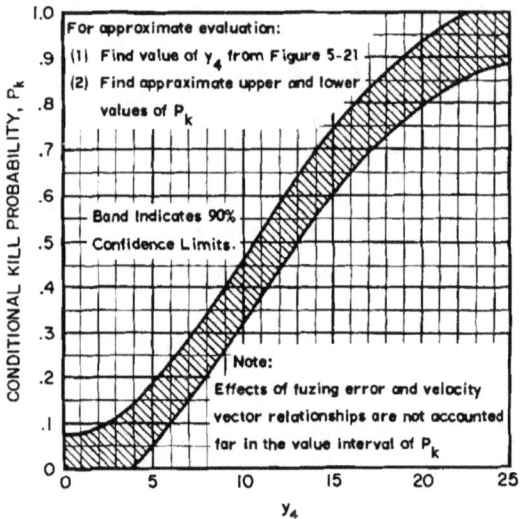

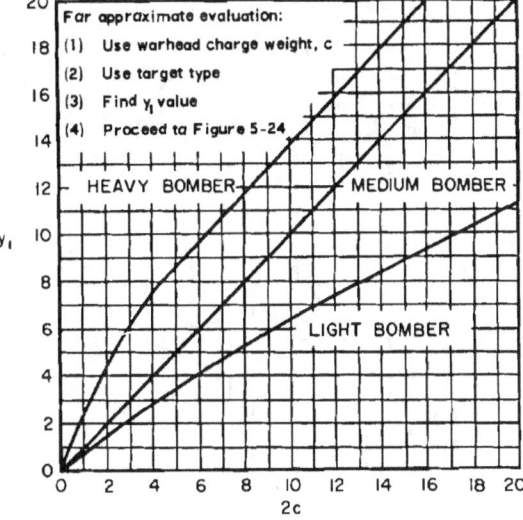

Figure 5-23. Internal Blast Warhead Evaluation-
Target Type Variable

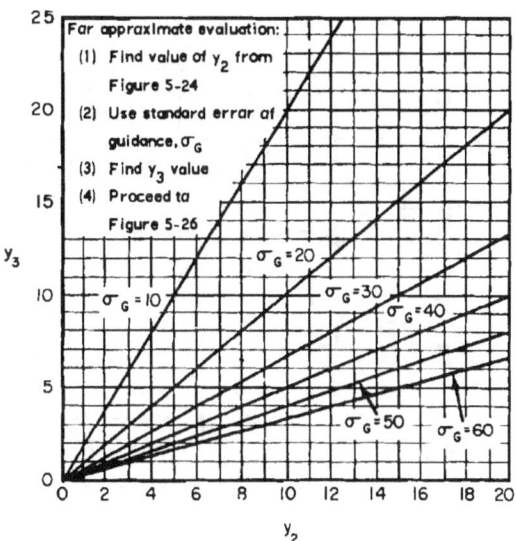

Figure 5-25. Internal Blast Warhead Evaluation-
Standard Error of Guidance Variable

Figure 5-24. Internal Blast Warhead Evaluation-
Engagement Altitude Variable

Figure 5-26. Internal Blast Warhead Evaluation-
Kill Probability Intervals

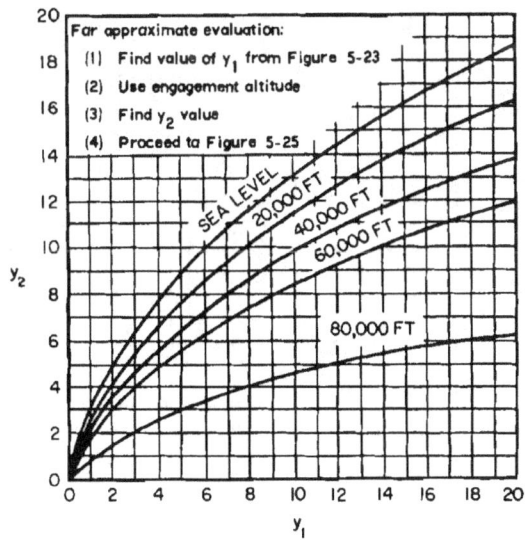

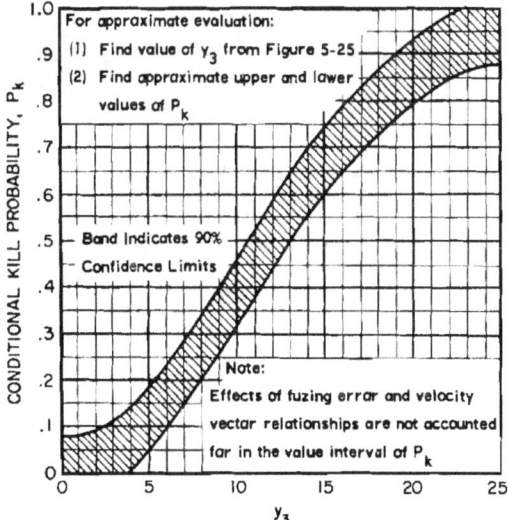

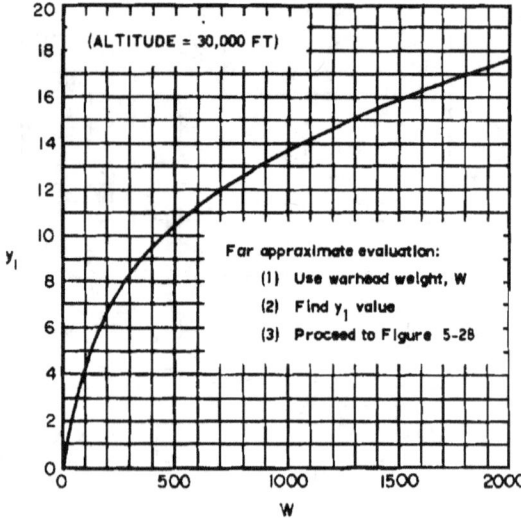

Figure 5-27. Fragment Warhead Evaluation - Warhead Weight Variable

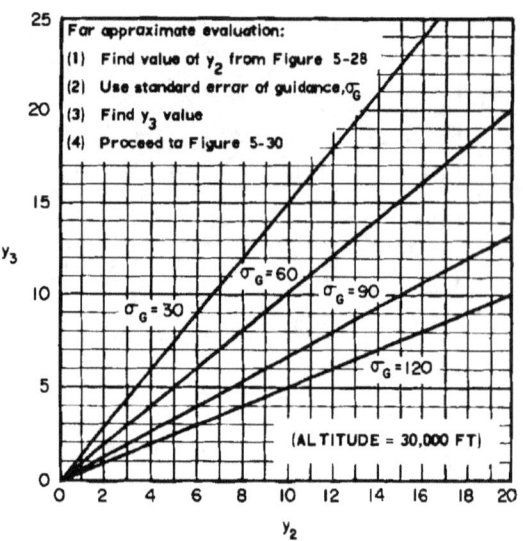

Figure 5-29. Fragment Warhead Evaluation - Standard Error of Guidance Variable

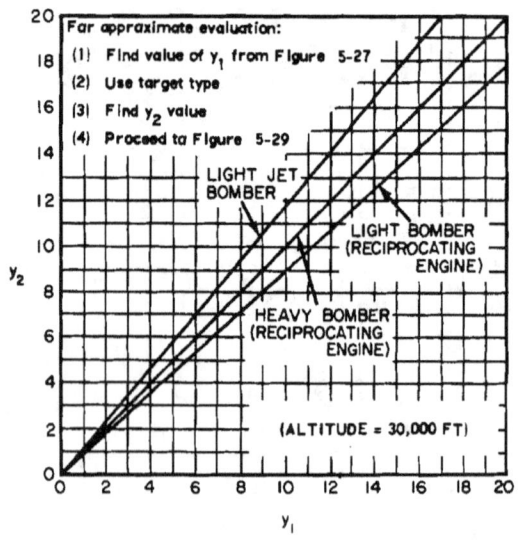

Figure 5-28. Fragment Warhead Evaluation - Target Type Variable

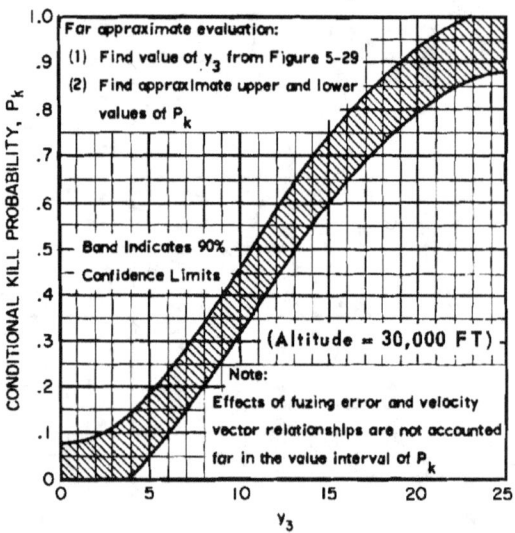

Figure 5-30. Fragment Warhead Evaluation Kill Probability Intervals

5-r "Vulnerability of F-84, F-86, F-94 and F-6U Jet Aircraft to Internal Blast", W. E. Baker, O. T. Johnson and R. T. Shanahan, BRL Report 848, October, 1954.

5-s "Vulnerability of Aircraft to Internal Blast", Irene M. Cooney, BRL Memo. Report 542, April, 1952.

5-t "Internal Blast Damage to Aircraft at High Altitude", BRL Memo. Report 605, April, 1952.

5-u "Vulnerability of B-47 Wing to Internal Blast", W. E. Baker and O. T. Johnson, BRL Memo. Report 531, March, 1951.

5-v "The Relative Internal Blast Vulnerability of Some Simulated Aircraft Interspar Wing Sections", W. E. Baker and O. T. Johnson, BRL Tech. Note 557, November, 1951.

5-w "Computation of Survival Probability of a Multi-Component Airplane", J. I. Brown, NAVORD 3597, February, 1954.

5-x "A Theory of Fragmentation", N. F. Mott and E. H. Linfort, January, 1943.

5-y "Justification of an Exponential Fall-Off Law for Number of Effective Fragments", H. K. Weiss, BRL Report 697, February, 1949.

5-z "Consideration of the Effects of the Size of a Projectile on the Efficiency of its Fragmentation", R. H. Kent, BRL Report X-58, February, 1933.

5-aa "Damage by Controlled Fragments to Aircraft and Aircraft Components", Arthur Stein and H. Kostiak, BRL Memo. Report 487, February, 1949.

5-bb "Damage to Aircraft and Aircraft Components by Fragments of Known Mass and Velocity from Controlled Fragmentation Shell", Johns Hopkins University Project THOR Technical Report 1, June, 1949.

5-cc "Vulnerable Areas of B-25 Pilot and Copilot", BRL Memo. Report 538.

5-dd "Single Shot Kill Probabilities of NIKE I for Two Fragment Sizes", Stanley Sacks, BRL Memo. Report 627, October, 1952.

5-ee "Vulnerability of Light, Two-Engine Bombers to Guided Missiles with 300 Pound Fragmentation Warhead", Ed S. Smith and C. S. Mynes, BRL Memo. Report 534, April, 1951.

5-ff "Vulnerability of Heavy Four-Engine Bombers to Guided Missiles Having 300 Pound Fragmentation or Blast Warheads", Ed S. Smith, C. S. Mynes, and W. Stubbs, BRL Memo. Report 540, May, 1951.

5-gg "Optimum Warheads and Burst Points for Bomarc, Phase II Guided Missiles", Ed S. Smith, A. K. Eittreim, and W. L. Stubbs, BRL Memo. Report 739, November, 1953.

5-hh "New Casualty Criteria for Wounding by Fragments", Allen and Sperrazza, BRL Report 996, October, 1956.

5-6. BIBLIOGRAPHY

(1) "The Monte Carlo Method and Its Applications", M. Donsker and M. Kac, International Business Machines Corp., New York, 1951.

(2) "Stochastic (Monte Carlo) Attenuation Analysis", H. Kahn, RAND Corporation Report R-163, 1949.

(3) "Methods of Reducing Sample Size in Monte Carlo Computations", A. W. Marshall, Journal Operations Research Society of America, Volume 1, Pages 263-278, 1953.

(4) "The Monte Carlo Method as a Natural Mode of Expression in Operations Research", Journal Operations Research Society of America, Volume 1, Pages 46-51, 1953.

(5) "The Monte Carlo Method", N. Metropolis and S. Ulam, Journal American Statistical Association, Volume 44, Pages 335-341, 1949.

(6) "Monte Carlo Method", National Bureau of Standards Applied Mathematics Series, Volume 12, 1951.

(7) "Modern Mathematics for the Engineer", E. F. Beckenbach, McGraw-Hill, New York, 1956.

(8) "Lotto Method of Computing Kill Probability of Large Warheads", F. G. King, BRL Memo. Report 530, ASTIA ATI-94269, December, 1950.

(9) "A Simple Method for Evaluating Blast Effects on Buildings", Armour Research Foundation, Chicago, ASTIA AD-38891, July, 1954.

(10) "On the Description of a Target Air-

craft", Robert W. Cross, Purdue University Statistical Laboratory Technical Note No. 1, ASTIA AD-57542, November, 1954.

(11) "Description of a Lethal Area Computation Problem", Herbert K. Weiss, BRL Memo. Report 723, ASTIA AD-21133, September, 1953.

(12) "Warhead Size and Effectiveness", H. H. Porter, Johns Hopkins University, Applied Physics Laboratory, CM-50, ASTIA ATI-32179, May, 1945.

(13) "Mathematical Methods of Statistics", H. Cramér, Princeton University Press, 1946.

(14) "An Introduction to Probability Theory and Its Applications", W. Feller, John Wiley and Sons, New York, 1950.

(15) "Statistical Theory with Engineering Applications", John Wiley and Sons, New York, 1952.

(16) "The Design and Analysis of Experiments", O. Kempthorne, John Wiley and Sons, New York, 1952.

(17) "Statistical Tables and Formulas", A. Hald, John Wiley and Sons, New York, 1952.

(18) "On the Collection and Handling of Data", Alfred N. Bock, Army Chemical Corps Manual No. 1, ASTIA AD-10830, May, 1952.

(19) "Probability Functions Associated With Measures of Vulnerability", J. R. Steinhilfer, Columbia Research and Development Corp. Report No. 124-4, ASTIA AD-82736, November, 1955.

(20) "Random Sampling Numbers", M. G. Kendall and B. B. Smith, Tracts for Computers, 2nd Series, No. XXIV, Cambridge University Press, London, 1939.

(21) "Mathematical Models In Large-Scale Computing Units", D. H. Lehmer, Proceedings Second Symposium on Large-Scale Digital Calculating Machinery, 1949, Pages 141-146, Harvard University Press, 1951.

(22) "Computation of Missile Trajectories in Three Dimensions", Clogett Bowie, Glenn L. Martin Co., January, 1957.

(23) "Methods for Computing the Effectiveness of Fragmentation Weapons Against Targets on the Ground", Herbert K. Weiss, BRL Report 800, January, 1952.

(24) "Methods for Computing the Effectiveness of Area Weapons", Herbert K. Weiss, BRL Report 879, September, 1953.

(25) "A Technique for Computing the Effectiveness of Fragmenting Warhead Missiles", F. S. Acton, Princeton University Technical Report 4, ASTIA AD-53412, Oct. 1953.

(26) "Methods Used at R.A.E. for the Assessment of the Lethality of Fragmenting Anti-Aircraft Munitions", J. K. S. Clayton and G. C. A. Raston, Royal Aircraft Establishment (Great Britain) Technical Note No. G. W. 372, ASTIA AD-70860, June, 1955.

(27) "A Standardized Procedure for Computing Vulnerable Areas Using I. B. M. Equipment", Donald Freeland, Purdue University Statistical Laboratory Technical Report No. 5, ASTIA AD-57547, September, 1954.

(28) "An Alternate Method for Computing Vulnerable Area", Nick Vaughan, Purdue University Statistical Laboratory Technical Report No. 3, ASTIA AD-57545, December, 1954.

(29) "A Mathematical Formulation for Ordvac Computation of the Probability of Kill of an Airplane by a Missile", M. L. Juncosa and D. M. Young, BRL Report 867, ASTIA AD-17267, May, 1953.

(30) "The Use of Gamma Rays for the Simulation of Warhead Fragment Damage", H. R. Crane, University of Michigan, ASTIA ATI-133952, December, 1951.

(31) "Aircraft Vulnerability as a Function of Fragmentation Penetration", C. F. Meyer, H. S. Morton and H. H. Porter, Johns Hopkins University, Applied Physics Laboratory, TG-24, ASTIA ATI-26719, April, 1947.

(32) "Effectiveness of Warheads for Guided Missiles Used Against Aircraft", Ed S. Smith, BRL Memo. Report 507, ASTIA ATI-75527, March, 1950.

(33) "A Study of the Effectiveness of a Mother-Daughter Type Warhead for the Sparrow Missile", Sperry Gyroscope Co., ASTIA ATI-152369, May, 1950.

(34) "Thumper Project-Study of the Prob-

ability of Destruction of Enemy Missile", H. Chestnut, General Electric Co., Report TR-45849, ASTIA ATI-2754, March, 1947.

(35) "Thumper Project - Approximate Methods for Determining Probability of Target Destruction", H. Chestnut, General Electric Co. Report TR-55302, ASTIA ATI 3892, April, 1947.

(36) "The Effectiveness of Zeus and Related Missiles", ASTIA ATI-76522, December, 1949.

(37) "Test and Evaluation of the XKD24-1 Target PA", George M. Miller, Naval Air Missile Test Center Technical Report 32, ASTIA ATI-36922, September, 1948.

(38) "Development and Evaluation Testing of KDH-1 Target PA", George M. Miller, Naval Air Missile Test Center Technical Report 45, ASTIA ATI-52642, April, 1949.

(39) "Report of Evaluation Panel on Bumblebee Project", Johns Hopkins University Applied Physics Laboratory, ASTIA ATI-108855, June, 1946.

(40) "Efficiency of Assumed Fuzings for Bomarc Phase II Guided Missiles", Ed S. Smith, et al, BRL Memo. Report 776, March, 1954.

(41) "Procedures for Obtaining Binomial Probabilities Within Three Decimal Accuracy Universally", Ed S. Smith, BRL Report 718, May, 1950.

(41a) "Binomial Normal & Poisson Probabilities", Ed S. Smith, Published and distributed by Ed S. Smith, Box 279, R. D. 2, Bel Air, Maryland, 1953. Revision of BRL Report 718.

(42) "Tables of Cumulative Binomial Probabilities", Ordnance Corps ORDP 20-1, September, 1952.

(43) "Theory of Probability with Applications", Henry Scheffe, NDRC A-224, OSRD No. 1918, 14 February 1944.

(44) "An Introduction to the Analysis of the Results of Firing Trials", T. R. Gemmell, CARDE Report No. 288/52, September, 1952.

(45) "Complex of Soviet Ground Targets on a Stabilized Front", Wm. A. McKean (Lt. Col. Inf.) and Ed S. Smith, BRL Memo. Report 855.

(46) "Poisson's Exponential Binomial Limit Tables", E. C. Molina, D. Van Nostrand Co., Inc., N.Y.C., 1942.

(47) "Offset Circle Hit Probabilities", RAND Tables 234.

(48) "1500 lb Antipersonnel Warhead for the Honest John Rocket", Ed S. Smith et al, BRL Memo. Report 779.

(49) "Salvo Hit Probabilities for Offset Circular Targets", A. D. Groves and Ed S. Smith, BRL Tech. Note 1088.

(50) "Effectiveness of Bomarc 300 lb Warheads Against B-29 Type Bombers", Ed S. Smith et al, BRL Memo. Report 595, January, 1952.

(51) "A Comparison of the Effectiveness of Conventional Rifles with an Experimental 'Salvo Weapon' ", Theodore E. Sterne, BRL Memo. Report No. 951, January, 1956 (Formulation for Salvo and determining σ from hits within a given radius).

(52) "Optimum Warheads and Burst Points for Bomarc, Phase II Guided Missiles", Ed S. Smith, A. K. Eittreim, W. L. Stubbs, BRL Memo. Report No. 739, November, 1953.

(53) "Multiple Bombing of Targets Having an Exponential Density Fall-Off", BRL Report No. 895, Charles E. Clark, and G. Trevor Williams, February, 1954.

(54) "Exposure to Airburst Warheads of Men in an Artillery Battery and in Infantry Positions", Ed S. Smith, BRL Memo. Report No. 1115, November, 1957. (Also see BRL Memo. Report No. 1067.)

(55) "Elementary Comparison of Antiaircraft Warhead Types", Herbert K. Weiss, BRL Memo. Report No. 631, November, 1952.

(56) "A Mathematical Formulation for ORDVAC Computation of the Probability of Kill of an Airplane by a Missile", M. L. Juncosa and D. M. Young, BRL Report 867, 1953.

(57) "Tables of Probability Density Function $kx^a e^{-bx^c}$ ", Charles E. Clark, BRL Report No. 1007, February, 1957.

(58) "Table of Salvo Kill Probabilities for Square Targets", National Bureau of Standards, Applied Mathematics Series, 44, 1954.

(59) "Expected Coverage of a Circular Target with a Salvo of n Area Kill Weapons", Arthur D. Groves, BRL Memo. Report No. 1084, July, 1957.

(60) "The Effectiveness of Various Weapons Used in Air Attack on Ground Troops", M. Trauring, BRL Report No. 754, May, 1951.

(61) "A Class of Casualty Functions with Special Application to Circular Targets", Sandia Corporation, Albuquerque, New Mexico, Case 417.000, August, 1954.

(62) "A Concept Armament System for the Main Battle Tank", D. C. Hardison and B. N. Goulet, BRL Tech. Note 1183, April, 1958.

Chapter 6
WARHEAD TESTING

6-1. INTRODUCTION

The warhead designer needs to know the philosophies and techniques associated with the testing of warheads for missiles. Testing and experimentation of some type is a basic requirement during research and development leading to the successful design of a guided missile warhead. Component testing is required to verify assumptions and to determine unknowns which cannot be predetermined analytically. With the exception of combat experience, it is only through testing that the effectiveness and reliability of a warhead and its weapon system can be defined. Therefore, consideration must be given by the warhead designer to the formulation and execution of an adequate test program which will produce the required data economically and at the proper time.

6-2. PLANNING OF TEST PROGRAM

6-2.1. Introduction The testing associated with the development of a new warhead is a complex operation involving the coordinated effort of many people, often over a considerable duration of time. The data generated by the test program is needed by the warhead designer at particular times during the warhead development program. Therefore, careful planning of the test program should be initiated as soon as possible after the design requirements for the warhead system have been established.

The approximate extent of testing required during the development period can be determined by clearly setting forth the objectives for the warhead and its associated components. A test program may then be planned to determine if these objectives can or have been achieved and to obtain basic information for the development of the warhead design. Since it is entirely possible that some of the required information is already available, it is recommended that all possible reference sources be thoroughly investigated by the warhead designer prior to planning the test program.

Once detailed development and testing has begun, additional test requirements will probably become apparent. In many instances, these new requirements will originate during the testing of the various warhead components. Budgeting for a test program should therefore be flexible enough to allow for such additional requirements.

6-2.2. Outlining the Test Requirements The warhead designer should first outline the basic problems that must be investigated in the development program. With this outline as a guide, the test requirements can be established and the necessary facilities, equipment, and instrumentation can be determined. A typical outline giving the basic parameters which the designer would investigate in a test program follows. A cluster warhead has been chosen as an example, since the problems which are associated with this warhead are representative of the other warhead types.

CLUSTER WARHEAD-
PARAMETERS OF INTEREST

I. COMPONENTS
   A. Submissile
      1. Aerodynamic Characteristics

    2. Damage Mechanism Characteristics
    3. Structural Integrity
    4. Number
  B. Submissile Ejection System
    1. Ejection Velocity
    2. Submissile Damage
    3. Adjacent Component Damage
  C. Skin or Fin Removal System
    1. System Effectiveness
    2. Adjacent Component Damage
    3. Ejection Sequence
  D. Warhead Structure
    1. Structural Integrity
  E. Initiation System

II. WARHEAD SYSTEM
  A. Component Functioning
    1. Static Conditions
    2. Dynamic Conditions
  B. Dispersion
    1. Static Conditions
    2. Dynamic Conditions
  C. Specified Engagement Conditions
    1. Potential Lethality
    2. Reliability

6-2.3. Establishing a Specific Test Program Once the general outline of the test program has been established, specific tests should be planned as far in the future as possible. Among the factors to be considered when planning for these tests are the availability of instrumentation, test facilities and ranges. Safety may also be a vital problem. Methods of reducing and analyzing the resulting data must be set up. Having reviewed his test requirements in view of the above factors, the designer should be able to decide on the most appropriate method of securing the required data.

Of primary importance in the planning of a test program is the determination of the type and availability of the instrumentation necessary to secure the desired information. Inherent reliability and ease of data reduction and analysis are major factors influencing the selection of instruments. Generally, as the instrumentation setup becomes more complex, the reliability decreases. This usually requires some duplication of instrumentation to assure sufficient data output. It may be necessary to spend some time developing new instrumentation if available equipment is unsatisfactory.

The availability of suitable test ranges must be determined and arrangements made for their use through appropriate channels. Instrumentation and trained personnel are available at most testing agencies. Organizations having extensive test facilities usually have a standard procedure for scheduling test programs. This scheduling is done through an office which assigns a priority to the program. The warhead designer should expect some delay during the scheduling phase. In addition, a sufficient amount of time should be allowed for pre-test preparations. This could involve a few hours for a simple test or many weeks for a more complex program. Certain programs may require the construction of a special test facility. In some instances, delay may be caused by lack of coordination within a test facility where several organizations are contributing to the effort. One unit may be loading the test item, a second building a special test fixture, a third fabricating the fuzing system, and still a fourth handling the instrumentation. Adverse weather conditions can also delay a program. After completion of the test, it may be costly and time consuming to dismantle the test setup.

An important factor to be considered in planning a test program is the availability of the test item. Developmental material is nearly always costly. Also, it is not always possible to duplicate or repeat tests. It is very important to consider every possible problem when planning a program to test a developmental item. Tests should be carefully planned and the instrumentation should be thoroughly checked before use.

Safety is a prime consideration and every effort should be made during the planning stage to insure a maximum consideration of safety before, during, and after the tests. Care must be taken during the design of the test item to insure that the detonators and initiators will be the last components installed in the test package. Personnel, the instrumentation and adjoining installations must be protected from stray fragments as well as blast damage. Testing of incendiary warheads must be conducted in a location which minimizes the fire hazard. Wherever a fire hazard is present, some type of fire fighting equipment should be present. The testing of CW and BW warheads presents special safety problems. All test facilities should have a safety officer who has the responsibility for reviewing and approving the test program, and who is to be consulted when there are special problems.

6-2.4. Data Reduction and Interpretation The instrumentation and test facilities are important factors that affect the early stages of planning a test; however, the warhead designer should not overlook the later problems associated with data reduction and interpretation. These often affect the selection of instrumentation and test facilities. Where data-reduction organizations are available at the test facility, it is advantageous to make use of them whenever possible.

The warhead designer will find that there are several limiting factors to an organization's ability to reduce and assimilate data. Sufficient trained personnel and equipment may not be available or may be too busy to take on additional work. If the organization is capable of accepting the work, there may still be the problems associated with the inherent conditions of the test. These include such items as poor film exposures and the use of instrumentation which is less than optimum. In many instances, the tested item does not function as expected. For instance, a missile may fall so far short of its target as to cause all of the cameras to fail to record the impact point. The conditions under which the test is conducted contribute to these problems. The dispersion of some types of warheads, such as fragment warheads, will cover a large area. This could mean that a considerable amount of time must be spent locating and plotting the fragment impact points. Bad terrain will compound the difficulties, although tests of this type are usually conducted over cleared areas. However, it is difficult to keep these areas completely clear of underbrush. Sometimes the wind will reverse, blowing the test items into wooded areas. Since many dispersion tests are conducted over sandy desert areas, it is possible for the test item to bury itself in the sand, making recovery difficult. The designer should inspect these ranges and inquire about any local conditions which could hinder data acquisition, such as rainfall or sandstorms. It may be necessary to conduct the tests under weather conditions less than ideal. This could result in poor film exposures; these create a particularly difficult problem if high speed photography is involved.

An important limiting factor in the taking of data is instrumentation failure. This is particularly a problem wherever a large amount of complex equipment is used. To prevent a complete loss of data, dual or alternate instrumentation should be used. For example, it is particularly important to duplicate high-speed cameras. The engineer, when planning for instrumentation, should anticipate emergencies so as to provide a sufficient number of alternate means for securing the data. Then, if there is an instrument malfunction, it may still be possible to reduce the remaining data. During many testing programs, there are repeated instrumentation malfunctions. When time and funding are important factors, the warhead designer should insist that the malfunctioning equipment be replaced. It is better to postpone a test and wait for adequate instrumentation, than to proceed with equipment which is not functioning properly. If there are continued data reduction problems because of poor instrumentation, the designer should investigate alternate means and approaches.

Weapon evaluation can be achieved by comparing the test results which the design objectives or with some previously established performance criteria. If these criteria are not available or are questionable, the warhead designer may either accept the test results or

6-3. TEST PROCEDURES AND TECHNIQUES

6-3.1. Introduction Warhead testing is divided into two phases. The first phase includes tests required to verify the effectiveness of the selected warhead damage mechanism. This is done before any large scale warhead designing or testing is begun. Parameters investigated include, for examples, lethality limits, penetration, and effects of blast. Testing of this type is conducted on fragments, rods and cluster submissiles.

The second phase includes testing of fractional or complete warheads. Tests of this type are conducted to investigate the operation and effectiveness of the warhead. They are usually conducted after the effectiveness of the

6-3.2. Phase 1 - Damage Mechanism Testing

Fragments Fragment lethality is determined by firing individual or small groups of fragments at typical targets. In many cases, these are the same targets which would be encountered in actual combat operations. When this is not feasible, simulated targets are used. Typical targets include aircraft sections, tanks, armor plate and other target materials. Gelatin blocks are sometimes used to simulate human targets. The consistency of gelatin approximates that of soft parts of the human body. These blocks are convenient since they can be molded to any size or shape. The results of

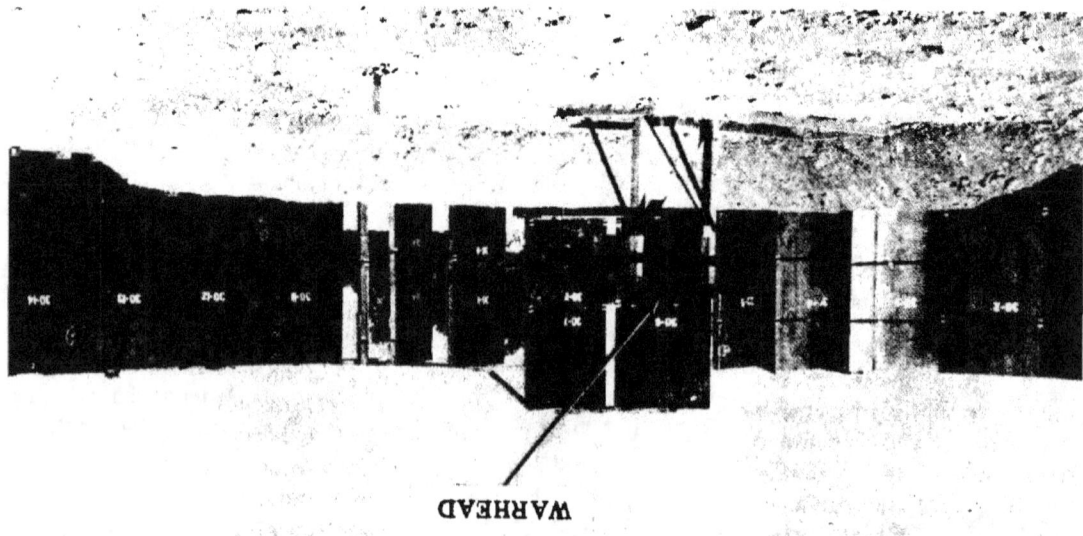

Aerojet Photo 955542

Steel plates and other target materials are arranged at various distances and orientations from the warhead and show fragment penetration and distribution.

Figure 6-1. Fragment Lethality Test

conduct a separate testing program to establish new criteria to verify the questionable performance criteria.

A description of the experimental techniques and procedures associated with each warhead type follows.

damage producing mechanism has been established.

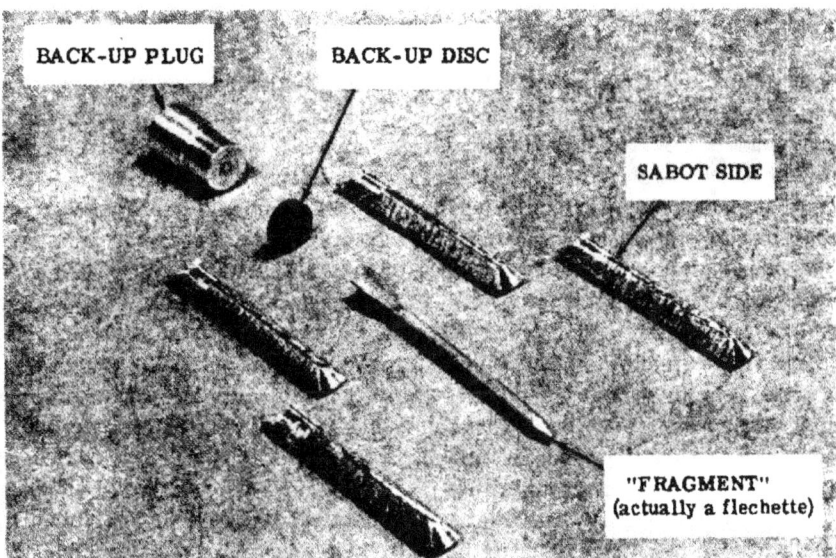

Exploded View - Fragment plus Sabot Components

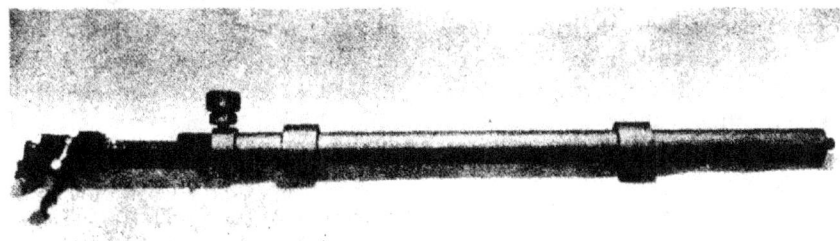

Barrel and Receiver - Fragment Gun

Figure 6-2. Fragment Gun and Sabot

firings are usually determined by visual observation of the blocks or photographs thereof.

Firing of fragments can usually be accomplished with a modified gun. The gun chosen should be capable of being fitted with a barrel sufficiently large to accommodate the fragments. It is necessary to contain the fragments being fired in sabots which so support the fragments that they may be properly accelerated in the gun. Sabots are usually designed to separate from the fragment when leaving the barrel. Once a firing program of this type is decided upon, the test engineer must select or design a suitable gun and cartridge case, and also design a sabot for the fragment being tested. It is then necessary to develop a powder charge which will impart the desired velocity to the fragment. It is unnecessary to waste sabots when developing the powder charge, since equivalent weight slugs of the correct bore diameter can be substituted for the sabot-fragment combination.

There are three methods by which fragment velocities can be measured. These are electronic, photographic, and penetration. Of the three, the electronic method is the most suitable for single firings. A counter chronograph is used to measure the fragment flight time between two screens which are a known distance apart. The screens are rigged in such a manner as to have the fragment close an electrical circuit when passing through the screen. A typical (make) screen would have two electrically conductive materials separated by an insulator. A fragment passing through the screen would short the two conductors, thus sending an impulse to the chronograph. As an alternate, the chronograph could be rigged to be triggered by having the fragment break an electrical circuit. The circuit in this instance could be a series of wires stretched across the fragments' flight path. These two methods are satisfactory when fragment accuracy cannot be guaranteed. If the screens are to be used over 100 feet from the chronograph, it may be necessary to use an amplifier in the line to boost the signal. When more accurate prediction of the fragment path is possible, lumaline screens can be used. These screens have photoelectric cells which give an impulse when a light beam is broken. The velocity can also be measured from high speed movies of the fragment in flight, although this method is more difficult since the equipment is not easily available and the data reduction can be time consuming.

Penetration into celotex is often used as indication of velocity. This is a crude method as penetration depth varies with both the angle of impact and the consistency of the celotex.

When the fragments are fired against simulated targets, penetration can be used as a criteria of the effectiveness of the fragments. This is especially true of aircraft structural sections, electronic equipment, vehicles, and infantry equipment. Infantry equipment used as targets includes helmets, armored vests, etc.

Cluster Submissiles Testing of the submissiles is required to determine their effectiveness in tactical use. Submissiles loaded with high explosives can be either the blast or fragmentation type. They are tested by being shot at or placed adjacent to obsolete aircraft, vehicles, equipment, and available components of new weapons likely to be targets. The submissiles are fired from a gun similar to that used for fragment lethality tests, in order to simulate the terminal velocity. The resulting information is best obtained by visual observation. Particular emphasis is placed on penetration, the amount of structural damage, the amount of equipment disabled, and after effects such as fires.

In some cases, fragmentation submissiles are tested for fragment lethality in the same manner as are fragment warheads. The setup does not have to be as large, although the same type of data is obtained (i.e., fragment size, weight, velocity, and distribution).

Testing is also required to determine the flight characteristics of both the stabilized and the unstabilized submissiles. This can be accomplished by testing a model of the submissile in a wind tunnel, firing it from a gun or dropping it from an airplane or tower. When the submissile is dropped, the flight characteristics are determined by tracking it with a phototheodolite, a conventional movie camera or a radar device. When the submissile is fired from a gun, it is necessary to determine both velocities and attitudes down range. Attitudes can be determined from microflash or shadow graphs, while flight attitude can be determined from holes in yaw screens. This is accomplished by examining the hole shape the test sample makes in a paper screen at a station along the line of flight. The submissile's velocity, if high, may be determined from the shadow graphs. If this is not feasible, velocities can be obtained using a counter chronograph and $n + 1$ screens for n velocity measurements required.

Rods The effectiveness against a typical target of a particular size rod must be known before it can be incorporated in a warhead. The warhead designer is therefore interested in establishing, first, the minimum degree of structural damage which must be inflicted upon a target to disable it, and secondly, in determining the size, material composition, and terminal velocity of the rod required to inflict this damage.

The degree of structural damage necessary to disable a target aircraft can be determined by a progressive artificial severing of the structural members in a typical aircraft

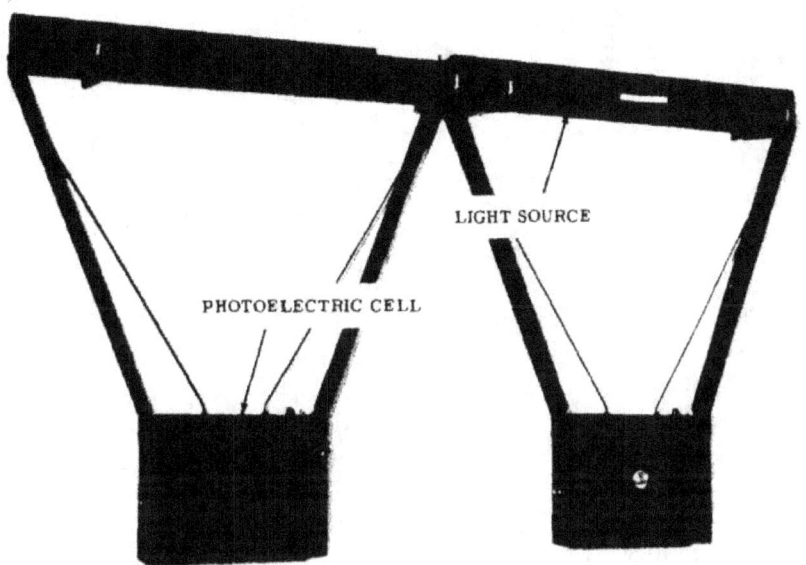

Lumaline screens containing light-sensitive cell which is activated when fragment passes through the triangular portion of the screen.

Figure 6-3. Screen Used for Measuring Velocity

section which has been placed under a simulated flight load. Thus, when failure of the section occurs, the amount of structure which a rod must cut to disable the target will be known.

The rod size, composition and velocity required to produce the necessary structural damage can be determined experimentally by observing the terminal ballistics of rods impacting against typical targets. Only by firing rods of various cross sections and materials into targets at various velocities and observing the resulting damage is it possible to determine the optimum rod configuration. For example, these targets may be scrapped aircraft sections or typical aircraft structural members. The rods are propelled by firing them from specially designed guns. In one application, the rods are supported in the barrel by a sabot which separates from the rod as it leaves the barrel. The procedures used to develop this type of gun are similar to those described in the section on fragments. When it is required that the rods hit the target with random orientation, a suitable object is placed in the rod's line of flight to cause the rod to tumble before striking the target. An alternate means of propelling the rod is to place it in a heavy metal support structure with some high explosive. The high explosive, when detonated, is contained in the metal structure so as to impart a high initial lateral velocity to the rod. (See Figure 6-4.)

A minimum of instrumentation is required for these tests to determine optimum rod configuration. It is first necessary to establish the rod velocities. This can be done either with high speed photography or (electronically) with a chronograph and screens. The terminal ballistics of the rods can be determined visually, by high speed photography or, if desired, with flash radiography.

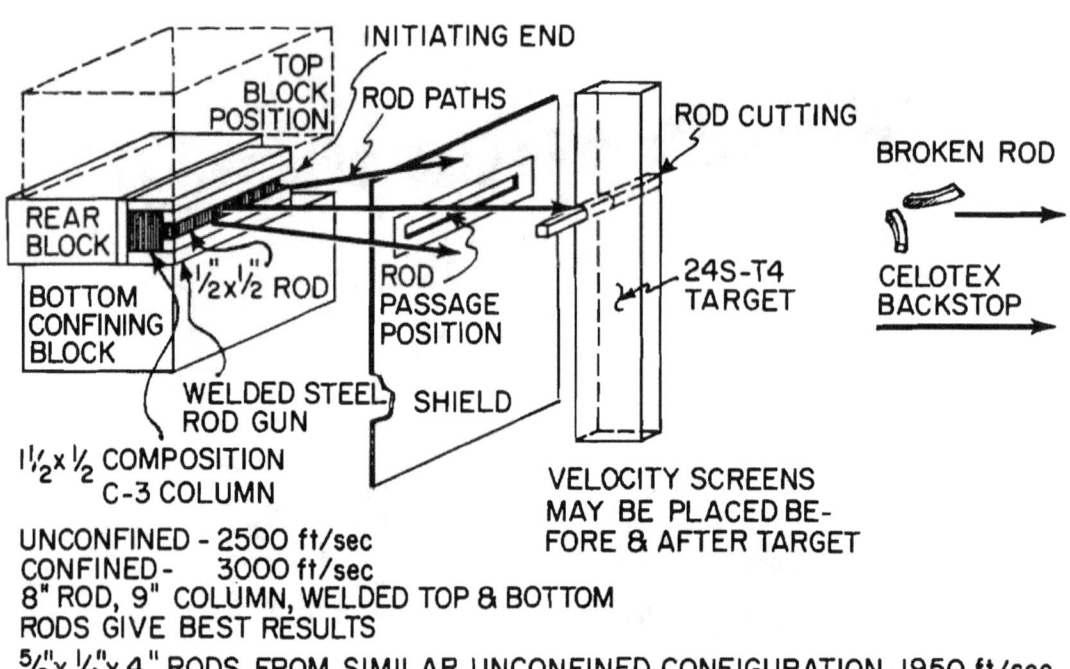

Figure 6-4. Individual Rod Test

6-3.3. Phase 11 - Warhead and Warhead Components

Fragmentation Warheads Two objectives are to be achieved when testing a fragmentation warhead. First, these tests confirm or deny the assumptions the designer has made, and secondly, they demonstrate the effectiveness and operation of the warhead design. The parameters of interest to the designer of a fragmentation warhead are the same regardless of the type of fragment used. These are the number, mass and distribution of potentially lethal fragments.

Most fragmentation warhead testing is done in circular or semicircular firing arenas with walls sufficiently thick to stop most of the fragments. Celotex is placed at suitable locations to insure a fair sampling of the fragments. The warhead is hung in the center of the arena on a gallows arrangement. As an alternate method, the warhead can be hung over a sand or water pit. Armored shelters are used to protect the men and instrumentation from blast or fragment damage.

The instrumentation used for the most part in these tests is photographic. Both high speed framing and smear cameras find applications for determining fragment velocity. Visual observation is sufficient for securing a large percentage of the data regarding dispersion and fragment breakup. The dispersion of the fragments is determined by plotting the impacts at different ranges with reference to a given point. When a warhead is detonated over a water pit, photographs are taken of the splashes caused by the fragments hitting the water. The blast effects may affect the clarity of the film in this kind of test.

Unstabilized fragments can be either preformed, such as spheres or cubes, or fireformed. Fire-formed fragments are formed by controlling the fragmentation of the warhead casing. Although the casing design may be theoretically correct, testing is required to confirm its fragmentation. Under certain circumstances, the test item may be scaled down to facilitate the testing procedure. By

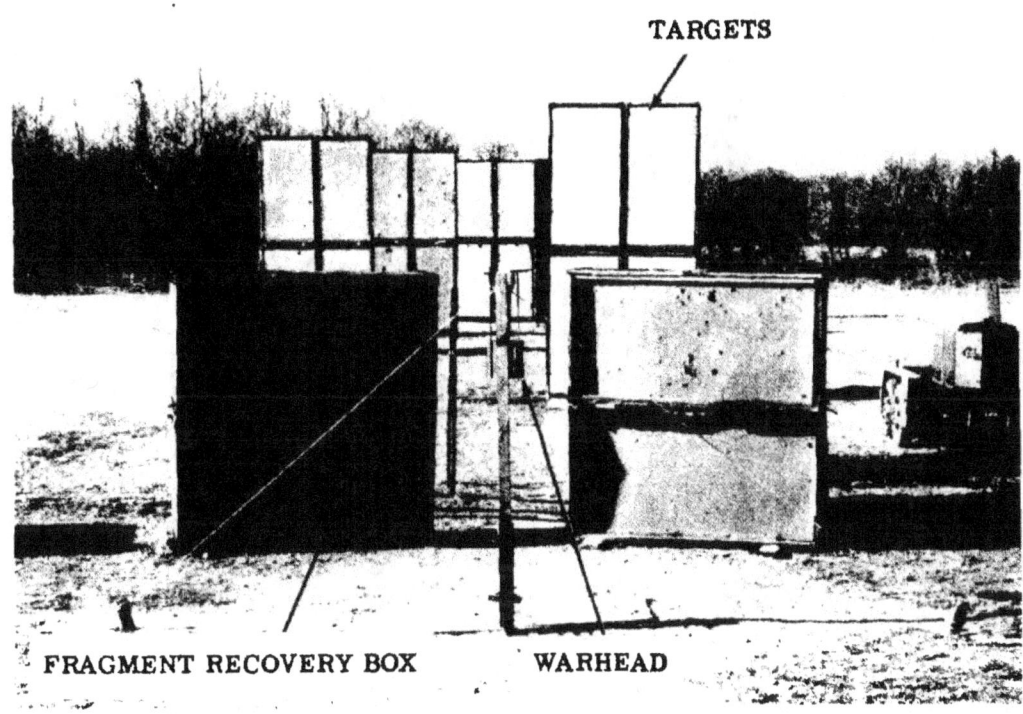

U. S. Army Photo

Field test set-up to determine down-range flight characteristics of fin stabilized fragments.

Figure 6-5. Complete Fragment Warhead Test

detonation over a sand pit, it is possible to recover the fragments and observe fragment breakup.

The initial fragment velocity depends primarily on the amount of high explosive used relative to the weight of fragment material. The fragment velocity can be measured by detonating scale model warheads having different charge-to-metal ratios and by photographing the fragments at a distance of ten to fifteen feet from the point of detonation with a smear camera. Smear cameras give, at a particular point, a time history which can be referenced to the time of detonation. The disadvantage of using smear cameras is their narrow width of field. When a large number of these warheads must be fired, it is feasible to simulate the fragment weight with inexpensive steel ballast on the parts of the warhead which will not be observed.

Fragment terminal velocity is often estimated by penetration of depth of penetration into celotex. When possible, high speed photographs of the fragments striking a target are taken. This is accomplished by choosing a target material which produces a spark when struck by the fragments. These sparks are photographed and the time is then related back to the initial detonation impulse. If times of flight to different ranges are known, then a velocity curve can be plotted. Another method of merit is photographing the fragments as they strike a frangible target. If these targets are sufficiently small, the location of the shattered targets provide a means of correlating dispersion, fragment shape and orientation with terminal velocity. It is sometimes feasible to color the fragments according to

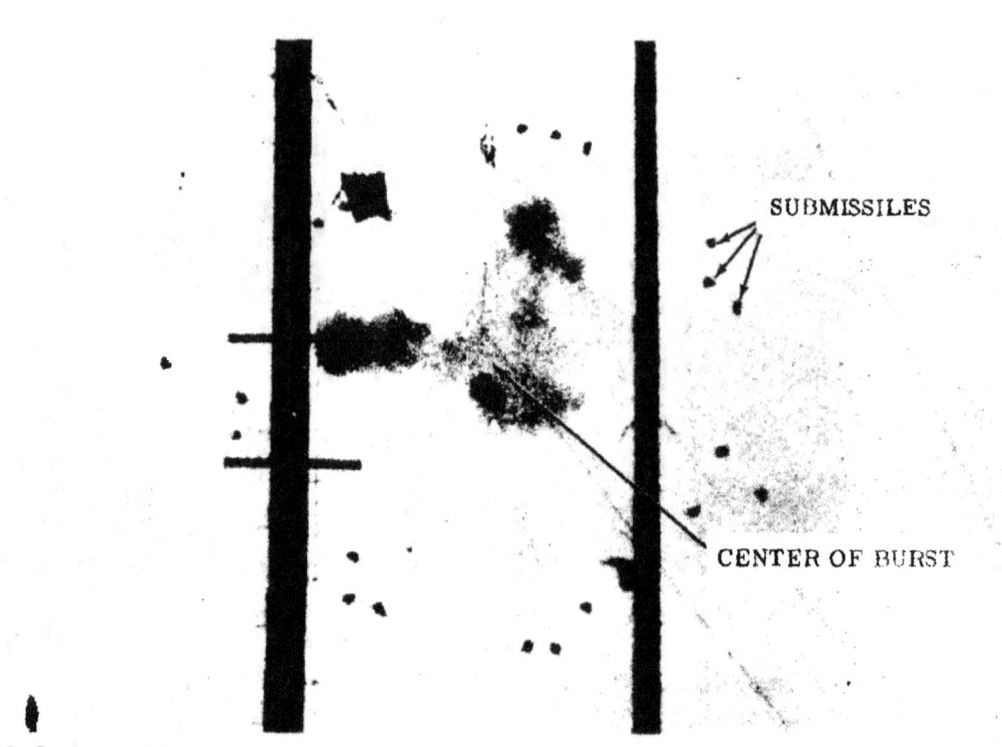

U. S. Army Photo

Warhead suspended between two poles with high-speed photographic coverage to determine submissile pattern.

Figure 6-6. Submissile Ejection System Test

their location in the warhead. This enables the project engineer to correlate data pertaining to recovered fragments with their original location in the warhead. (Ceramic paint must be used for this purpose since fragments get so hot when fired that their steel "blues".)

Cluster Warheads Because of the inherent complexity of the cluster warhead, the designer should anticipate an extensive test program. Consideration should be given to the different procedures and techniques required for the testing of cluster warheads and their components.

The effectiveness of a cluster warhead is dependent directly upon the successful functioning of its components. These components include the submissiles, submissile ejection system, envelope removal system, and structure.

There are two types of submissile ejection systems. When the submissiles must be forcibly ejected, a pyrotechnic system is used. Aerodynamic means are used when forcible ejection is not desired.

The objective of forcible ejection is to eject a submissile of given weight from the warhead at a specified velocity and with a specified maximum acceleration. The development of any such system requires an extensive amount of testing. The engineer should select a test site which affords a maximum of safety for the ejection method chosen. If a powder charge is selected for propulsion, instrumentation is required to determine chamber pressure versus time, thrust versus time, and the

Gun tube ejection type Warhead mounted on test sled — — view taken after firing shows the missile skin removed.

Figure 6-7. Skin Removal Test

missile ejection velocity. A reusable test stand is required to mount the submissile and ejection device. Pressure measurements are necessary to determine thrust as well as to insure that the ejection device has structural integrity. These measurements are made with either a transducer or a piezoelectric crystal gauge whose output is fed into an oscilloscope and recorded with a camera. Peak pressure can be determined with a copper crusher gauge. Thrust is measured with a strain gauge. Velocities are measured either photographically or with a counter chronograph, if the screens or circuits can be set up.

Once a powder charge has been developed, testing is still required to determine if all the submissiles can be ejected uniformly. This testing requires the use of an assembled warhead. Photographic coverage of the ejection sequence gives a maximum of information. A timing setup can be provided to determine the time from initial impulse to the time the last submissile is ejected. Visual observation will indicate whether the operation of the ejection system affects any of the other warhead components.

An aerodynamic ejection system relies primarily on aerodynamic forces to eject the submissiles after the warhead skin has been removed. During flight testing to determine the effectiveness of the system, photographic coverage is used to investigate functional performance of the system and any interference of other system components with the ejection sequence.

Skin ejection is accomplished explosively

178

with either shaped charges or detonating cord while fins are removed by explosive bolts. The major points are the effectiveness of the skin removal system and the effects of the function-ing of this sytem on warhead components. The test stand required for this testing is relatively simple; the primary problem is to prevent the stand from interfering with the detonation of explosive devices. Practically no instrumenta-tion is required for these tests since visual observation is adequate. Samples of the skin are placed on the stand, the removal device detonated, and the results noted. Where the effects on the other warhead components are required, these components, or simulated components, are oriented properly relative to the explosive. The charge is then detonated and the results noted.

The test procedure for a fin removal system is similar to that used for the skin removal system. The use of explosive bolts necessitates some additional testing to deter-mine the optimum bolt configuration. This involves functional testing of various bolt designs under static conditions.

It may be necessary to test the basic warhead structure to determine its strength under static and dynamic loads. These tests are performed in the conventional manner, and, if possible, the designer should consider testing the item to destruction.

The testing of the complete warhead system can be either static or dynamic. It is recommended to the designer that dynamic type testing be selected whenever possible. There will usually be two test objectives. The

High-speed camera coverage of cluster warhead mounted on test sled, showing ejection sequence.

Figure 6-8. Sled Test - Complete Cluster Warhead

U. S. Navy Photo NP/45-39870

Ice target prepared for test. Shaped charge is fired on stand near right side of picture so that all jet fragments are caught in the ice.

Figure 6-9. Complete Shaped Charge Warhead Test

first objective is to determine if the warhead functions as it was designed to, and the second is to observe the submissile pattern. The detonation of a cluster warhead under dynamic conditions provides the warhead project engineer with a fairly accurate picture of the performance of the warhead under actual conditions. A cluster warhead can be tested dynamically be a sled test, flight test, or air drop.

The selection of a sled or track facility is affected by the chance of damaging the track when detonating the warhead. By securing a track test facility, one can be certain that a maximum of information will be derived. Instrumentation for a test of this type can be as complex as the warhead designer requires. Generally, this includes extensive high speed photographic coverage to indicate the effectiveness of the skin removal system and the ejection system. A test of this type requires that the cameras be located accurately to insure proper coverage.

Flight testing or air dropping a warhead system is not recommended unless the components of the system have been thoroughly checked out and no other means were feasible. The instrumentation used in this case is generally photographic. Cameras are often located in chase aircraft as well as on the ground. Phototheodolites are used to determine the orientation of the missile before detonation. If the ground pattern is desired, the submissiles are ejected over a terrain which will facilitate their recovery.

Static testing of cluster warheads requires a test stand or tower. When results are analyzed, allowances may be made for interference with the submissiles by the tower structure. Again, instrumentation is photographic.

U. S. Army Photo

Fighter aircraft suspended between two towers with blast warhead detonated beneath it to determine warhead effectiveness.

Figure 6-10. Blast Warhead Test

Shaped Charge Warheads The warhead designer relies heavily on the results of previous research when designing a shaped charge warhead. The bulk of such results are obtained from experimental work. However, it is not possible, in most instances, to scale the results up or down to predict the performance of a new warhead. This is especially true for those warheads which are associated with missiles having a high rate of spin. Therefore, the warhead designer must perform a considerable amount of testing on full size warheads.

Basic investigations into shaped charge performance include testing to determine the effects of various charge parameters upon such factors as target penetration, hole volume, jet velocity, composition and shape. These tests are conducted either outdoors on a firing range or in a specially constructed test chamber. Dependent on the information required, the charges are detonated statically or dynamically. When it is desired to fire the charge from a gun for dynamic tests, special test cases are fabricated. These cases are used not only to contain the charge, but also to control the charge parameters so that a complete analysis can be made of the firing. When being tested statically, the charges can either be contained in the same type of case or tested in an unconfined condition, dependent on the information desired. The charge is usually either placed on a simple stand or hung, adjacent to the target at the proper standoff. Where standoffs of any large distance are used, it is necessary to hang the charge with some accuracy to insure hitting the target. The targets used for these tests range from the typical targets encountered in combat to material samples such as armor plate, mild steel, aluminum, reinforced concrete, large thicknesses of earth, and combinations thereof.

The data resulting from these tests can be secured by diverse methods including visual observation to determine penetration, and flash radiography to obtain data such as jet velocity, composition and shape. High speed photography also finds applications in this field. Examination of the warhead residue gives an indication of the operational efficiency of the warhead. The recovery of scorched, undetonated "chunks" of explosive, for instance, is an indication of a low order explosion. These tests and design modifications continue until a substantially optimum warhead configuration has been secured.

Small warheads can be fired from guns at specially prepared targets while large warheads must be installed in missiles for firing.

A typical target which could be located on the missile test range is a fortified combat communications center. It would be instrumented by placing simulated personnel at various locations in and around the target. Photographic coverage would be set up at a safe distance from the point of impact. Most of the data resulting from a firing of this type would be obtained by visual observation. The condition of the simulated humans would give an idea of the lethal radius of the warhead. The warhead debris would be examined for indications of malfunction. The presence of cone fragments, for example, would indicate that a shaped charge warhead did not detonate properly and would indicate a malfunction in, e.g., the fuzing system.

Rod Warheads Testing of a complete rod warhead usually does not begin until the optimum rod dimensions have been determined by evaluating individual rod firings. Complete warhead testing is initiated at an early stage in the development of the warhead since the resulting data governs the final configuration. This is of importance because most of the factors affecting the performance of a rod warhead are dependent upon the physical characteristics of the warhead. For instance, rod velocities and patterns are dependent upon the configuration of the explosive cavity. Rod velocities are also dependent upon the amount of high explosive used. Therefore, the warhead designer usually has various warhead configurations of a particular type fired before attaining a satisfactory design.

The test procedures for a rod warhead are very similar to those which are used for a complete fragmentation warhead. The warhead is hung or supported in the center of a test arena with targets placed around the circumference. Data which the warhead designer requires includes rod velocity, patterns, shapes, orientation and, for discrete rod warheads, dispersion. Target damage is also of interest.

Rod velocities are obtained by the use of high speed photography. Vertical aluminum tubes are placed in the ground at various distances from the point of detonation. When these tubes are struck by the rods, a bright flash is given off. This flash is picked up by the cameras and the impact time is correlated with timing marks on the film.

The rod pattern for a continuous rod warhead is obtained by visual observation. Thin sheets of metal, called pattern sheets, are placed perpendicular to the rod flight path at various distances from the point of detonation. The continous rods penetrate these sheets and leave an impression of the rod pattern.

Rod shapes and rod orientation can also be determined by placing wire screens perpendicular to the rod flight path at the particular ranges. When the rods strike the screen wire marks are impressed upon them. By observing these marks, one can determine the rod shape and orientation when the screen was struck.

It is sometimes feasible when conducting tests of this type to place aircraft structures in the arena. Thus, the lethality of the warhead can be determined.

After the warhead has reached an advanced stage of development, the designer can consider firing against a target drone. This test gives the closest determination, other than from actual combat, of the effectiveness of the warhead when detonated within proper range of the target. Instrumentation for this type of test is photographic. Cameras are located on the ground and possibly in chase aircraft. However, coverage is not good because of the extreme distances involved. It is sometimes possible to determine the amound of damage inflicted by visual observation of the downed target.

Blast Warheads The test procedures for a blast warhead are not of a complex nature, because of the relative simplicity of this type of warhead. The primary performance factor is the amount of drastic blast damage inflicted on a target. This is usually determined by detonating small warheads adjacent to different parts of the target and observing the resulting damage. Large warheads are detonated at varying distances from different parts of the target until the critical, lethal distance is found. With the larger warheads, blast pressures are sometimes measured at different distances from the point of detonation. Photography can be used for all of these tests to supplement the data secured by visual observation.

Additional variations in blast effect occur in tests where the target is near the ground. The enhancement due to ground approaches a limit of 100 per cent on a weight basis, as compared to tests in which the burst is high enough above the ground to be free from effects of the reflected blast wave.

6-4. DATA REDUCTION AND INTERPRETATION

Most of the instrumentation used in the testing of warheads does not present any serious data reduction problems. In many instances, data reduction is not necessary since the data obtained by visual observation is in finished form. These include such items as damage estimates, rod shapes and fragment dispersion. Data reduction is usually required when photographic and electronic instrumentation is used. Photographic instrumentation is used to a large extent in warhead testing for measuring fragment and rod velocities, shaped charge jet velocities, and cluster dispersion. Occasionally, the warhead designer will find applications for electronic instrumentation which, for warhead testing purposes, will require a minimum of data reduction effort. However, most of the raw data which the warhead designer handles is in the form of film.

High speed motion picture film can be analyzed frame by frame when it is necessary to determine the time at which an event occurred. This event could be a fragment striking a surface or a rod striking a flash tube. Film is read on film-viewers which project an enlarged image of the film, frame by frame, on a ground glass screen. The film reader can then correlate the frame in which the event occurred with a time signal imprinted on the film. Reading film appears to be a relatively easy task; however, it can be a difficult and tedious job if the film is not clear or if there is a large amount of data to be reduced.

An additional difficulty may be malfunctioning timing circuits. This can result in the complete or partial disappearance of the timing marks. Also, they may become lost in the background on the film. When the appearance of these timing marks is inconsistent or several of the marks are missing, it is possible to estimate the time by averaging the time intervals of adjacent frames.

High speed motion pictures can also be analyzed on a motion picture projector. This method is used when it is desired to observe in detail the performance of components such as skin ejection systems and submissile ejection systems. Data reduction in these cases is relatively easy. Movie film can be examined using a variable speed projector which can be stopped or reversed. The designer can thus determine exactly how the system in question is functioning. The acceleration and velocity of the submissiles can also be determined in this manner. The data reduction in this case is simplified if the motion picture is photographed against a calibrated background.

When evaluating the results of warhead and warhead component tests there are several factors which the warhead designer must consider. These include the data accuracy, test environment, scale effects and the quantity of

Table 6-1. Test Facility Selection Chart.

Warhead Static Tests / WARHEAD	ARMY					NAVY				AIR FORCE		
	ABERDEEN	PICATINNY	WHITE SANDS	DUGWAY	EDGEWOOD	NOL	NOTS	NPG	NAOTS	AFAC	AFFTC	HADC
FRAGMENTATION	X	X				X	X					
ROD	X	X										
CLUSTER	X	X										
SHAPED CHARGE	X	X				X	X	X				
HIGH EXPLOSIVE	X	X				X	X	X				
BW				X	X							
CW				X	X							
PROPAGANDA	X					X						

Warhead Dynamic Tests / WARHEAD	ARMY					NAVY				AIR FORCE		
	ABERDEEN	PICATINNY	WHITE SANDS	DUGWAY	EDGEWOOD	NOL	NOTS	NPG	NAOTS	AFAC	AFFTC	HADC
FRAGMENTATION			X			X	X	X			X	X
ROD			X			X	X	X			X	X
CLUSTER			X			X	X	X	X		X	X
SHAPED CHARGE	X		X			X	X	X				X
HIGH EXPLOSIVE	X		X			X	X	X				X
BW				X			X					
CW				X			X					
PROPAGANDA							X	X	X		X	

Damage Mechanism Tests / DAMAGE MECHANISM	ARMY					NAVY				AIR FORCE		
	ABERDEEN	PICATINNY	WHITE SANDS	DUGWAY	EDGEWOOD	NOL	NOTS	NPG	NAOTS	AFAC	AFFTC	HADC
FRAGMENTS	X	X				X	X					
RODS	X	X				X	X					
SHAPED CHARGE	X	X				X						
HIGH EXPLOSIVE	X	X										
BW				X	X							
CW				X	X							

data. Data accuracy is dependent upon the instrumentation used to obtain the data and the accuracy with which the data is reduced. The accuracy of the instrumentation and data reduction should be established both before and after the test is conducted. However, where accuracy is questionable, the engineer must determine the extent of the inaccuracy, qualify the evaluation or, if conditions permit, repeat the test.

The effects of the test environment should also be considered before testing is initiated. Typical environmental effects which may have to be considered include such items as the physical effects of test stands and structures, interference from the instrumentation, and interference from nearby warhead and missile components. Atmospheric conditions are also of concern.

When warheads and their components must be scaled up or down, there is always the problem of evaluating scale effects.

The warhead engineer is also faced with the possibility of instrumentation malfunction during a test program. When malfunctions occur, the quantity of data is diminished accordingly. Where there is a lack of data, it is sometimes possible to extrapolate the results. However, any conclusions based on such extrapolated data should be carefully qualified.

It is sometimes necessary to experimentally establish the criteria against which test results can be compared before a performance evaluation can be made. This does not present a problem when the development objectives are known. However, a criteria problem exists if the required velocities and accelerations are unknown. It is then necessary to establish the requirements experimentally. It should be noted that valuable design data is generated from any properly conducted test, and this data should be utilized to its fullest extent. Lethality is another criterion which is not always firmly established, and consequently may require experimental work.

The comparison of the test results with theoretically or experimentally established criteria is sufficient in most cases. However, the most positive way of determining the effectiveness of a warhead, outside of actual combat firing, is to fire the test item in its missile under simulated combat conditions. It should be remembered that the overall kill probability includes P_c (see subchapter 5-1) and other

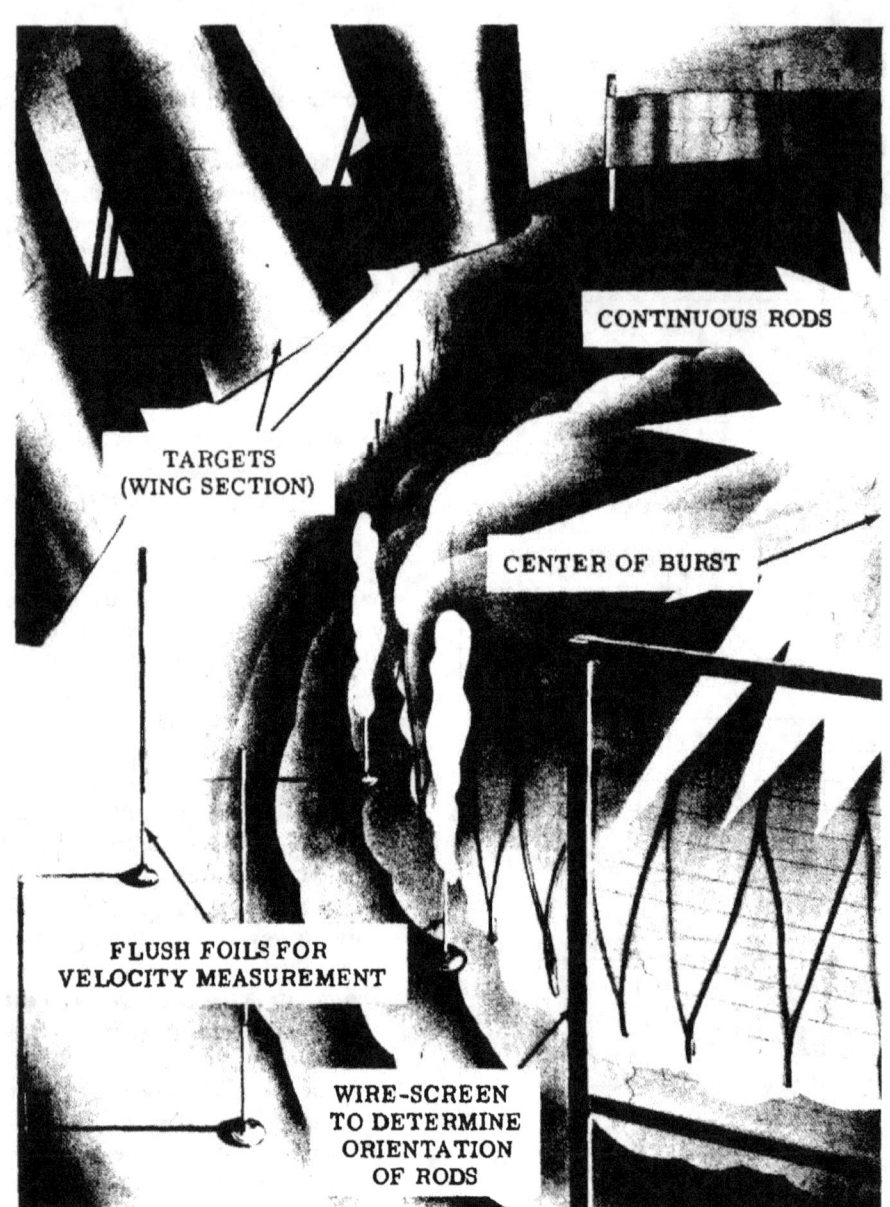

New Mexico Institute of Mining and Technology Photo

Figure 6-11. Complete Rod Warhead Test

factors which are, in part, dependent on the competence of the firing unit. Consequently, troop training must overlap prototype missile development.

6-5. TEST FACILITIES

This subchapter includes a description of some of the Government facilities available for the testing of warheads and their associated components. It is presented to permit the designer to select the appropriate installation at which the required testing can be conducted.

Test facility selection charts are presented in Table 6-1 as an aid. The following organizations are covered:

Department of the Army

    Ordnance Corps
        Aberdeen Proving Ground
        Picatinny Arsenal
        White Sands Missile Range
    Chemical Corps
        Army Chemical Center
        Dugway Proving Ground

Department of the Navy

    Bureau of Ordnance
        Naval Ordnance Laboratory
        Naval Ordnance Test Station
        Naval Proving Ground
        Naval Aviation Ordnance Test Station

Department of the Air Force

    Air Research and Development Command
        Air Proving Ground Center
        Air Force Flight Test Center
        Holloman Air Development Center

6-5.1. Aberdeen Proving Ground, Aberdeen, Maryland

The Aberdeen Proving Ground is the principal engineering and service testing center for Ordnance Corps equipment to be used by the Army Field Forces. It has the responsibility for determining the operational and functional ability of new Ordnance weapons and of equipment to be approved for production. The research and development mission of the Aberdeen Proving Ground is to carry out studies and experimental testing for the purpose of producing design criteria and for developing all types of weapons and instrumentation as required. The following material list of research and development Laboratories and other test facilities located at the Aberdeen Proving Ground is of interest to the warhead designer.

(1) Exterior Ballistics
    (a) Wind Tunnels
    (b) Free Flight Firing Ranges
    (c) Controlled Pressure-Temperature Ranges

(2) Interior Ballistics

(3) Terminal Ballistics
    (a) Shaped Charge Laboratory
    (b) Shaped Charge Firing Barricades
    (c) Blast Facilities
    (d) Shock Tubes
    (e) Fragmentation Chamber
    (f) Fragment Gun Range
    (g) High Altitude Facility
    (h) High Speed Ballistic Track

(4) Computing Laboratory

(5) Small Arms Range

(6) Major Caliber Range

(7) Armor and Armor Defeating Ammunition Testing
    (a) Armor Plate Ranges
    (b) Projectile Ranges
    (c) Armor and Ammunition Test Areas
    (d) Tank Vulnerability

(8) Aircraft Vulnerability and Ammunition Effectiveness Ranges

(9) Fragmentation Test Facility
    (a) Open Fragmentation Pits
    (b) Fragmentation Panels
    (c) Closed Fragmentation Pit
    (d) Fragment Velocity Measuring Instrumentation

(10) Bomb Testing Facilities

(11) Environmental Testing Facility

(12) Weapons Systems Evaluations

(13) Data Reduction

(14) Rocket Flight Testing Facilities

In addition to these test facilities, there are sufficient range areas available to provide for many special testing requirements of the warhead designer.

6-5.2. Picatinny Arsenal, Dover, New Jersey

The research and development mission of Picatinny Arsenal is to develop various types of munitions, including missile warheads. Among the test facilities located at this installation of interest to the warhead designer are the following:

- Instrumentation Development Laboratories
- High Acceleration Air Gun
- Wind Tunnels
- Fragmentation Chambers
- Ballistic Ranges
- Static Test Chambers
- Sectioning and Disassembly Chambers
- Explosive and Propellant Evaluation Facilities
- Rocket Testing Area
- Centrifugal Force and Vibration Equipment
- Ammunition Component Packaging and Handling Equipment (Design and Test)
- Plastic Research Test Equipment

These facilities may be used to check fragment, submissile or rod flight characteristics, to determine the characteristics of various high explosives, or to determine the effectiveness of scaled or full size warheads.

6-5.3. White Sands Missile Range, Las Cruces, New Mexico

The White Sands Missile Range is the principal Army Ordnance Corps installation for the execution of all technical and engineering responsibilities associated with the flight testing of guided missiles and other munitions. The Range is a joint service installation operated and administered by the Ordnance Corps for the three military departments. Flight and non-flight evaluation tests are conducted for engineering assessment, acceptance, user evaluation of ordnance (contractor developed), or other services' missiles.

The Army research and development mission at White Sands Missile Range is to:

(1) Prepare engineering test criteria and procedures.

(2) Provide technical facilities and operating personnel.

(3) Provide supporting services as required.

(4) Plan for, recommend and, where directed, provide special facilities and/or areas for testing material other than guided missiles and long range rockets.

(5) Conduct scientific investigations as required.

As the primary mission of this installation is to flight test guided missiles, the majority of its facilities have been created to support this objective. The available ranges can be utilized by the designer to test the warhead as part of a check-out of the overall missile system. This may include an investigation into the warhead terminal ballistics against assorted targets. It is also the mission of WSMR to test components and subsystems of guided missiles for overall evaluation when flight testing is not required.

6-5.4. Army Chemical Center, Edgewood, Maryland

The Army Chemical Center engages in basic and applied research and development and field testing. The Center conducts basic physiological, biochemical, and toxicological research on chemical warfare agents. It conducts research on wound ballistics for the purpose of improving both the defensive effectiveness of body armor and the offensive effectiveness of anti-personnel weapons. The Center has the responsibility for developing new chemical (and radiological) warfare agents, materials, and methods for both offensive and defensive purposes. In addition to these missions, the Center has the responsibility for conducting, evaluating, and preparing reports on engineering, field, and user tests of Chemical Corps material. These tests may be conducted in conjunction with other development and test agencies.

6-5.5. Dugway Proving Ground, Tooele, Utah

The Dugway Proving Ground is a combined research and test installation with the following missions:

(1) To obtain basic scientific data on new and improved CW, BW and RW agents.

(2) To conduct controlled field tests of CW, BW and RW agents and agent vehicles.

6-5.6. Naval Ordnance Laboratory, White Oak, Maryland

The Naval Ordnance Laboratory is a research and development organization that has the primary objective of originating and testing new devices in Naval Ordnance. Of primary interest to the warhead designer are its test facilities that include extensive aeroballistics and high explosives research areas. These areas can be utilized to perform basic research to determine both the characteristics of different high explosives and the flight characteristics of various shaped fragments. There are also additional range facilities available for warhead ground tests which do not require extensive range areas. In addition to these, there are the usual support facilities available, including various shops, and instrumentation and data reduction facilities.

6-5.7. Naval Ordnance Test Station, China Lake, California

The Naval Ordnance Test Station is engaged in research, effectiveness and feasibility investigations, design, development, product and production engineering, test and technical evaluation and the pilot production of ordnance materials, components, assemblies and systems in the field of missiles and other ordnance items. Facilities are available for research and development in the fields of high explosives and aeroballistics. Noteworthy test facilities include a moving-target range and several high speed tracks. Such supporting functions as instrumentation, data reduction, and H. E. handling facilities staffed with trained personnel are also available. In particular, the facilities can be used to secure data on warhead terminal ballistics, fragment flight characteristics, cluster warhead functioning under static or dynamic conditions, and the characteristics of high explosives. Sufficient range space is also available for the testing of shaped charge, blast, and other type warheads.

6-5.8. Naval Proving Ground, Dahlgren, Virginia

The mission of the Naval Proving Ground is to conduct proof tests of ordnance materiel, conduct research and development of ammunition and components thereof, armament and components thereof, armor and ballistics, etc., and to investigate other ordnance problems. Whenever necessary, research is also conducted to develop required instrumentation. The test facilities, including photographic and electronic instrumentation, are available for explosives-handling tests and interior ballistics studies. Ranges available include fragmentation arenas, indoor and outdoor firing ranges, and aerial gunnery and bombing ranges. These range facilities can be used to secure terminal ballistics data for blast, shaped charge, rod and fragment warheads. It is also possible to test full size (as well as scale model) fragmentation and rod warheads in order to secure data on such parameters as dispersion, fragment and rod size and shape, and fragment and rod velocities.

6-5.9. Naval Aviation Ordnance Test Station, Chincoteague, Virginia

The Naval Aviation Ordnance Test Station provides technical support and services to Government agencies and contractors in research, development, test, and evaluation programs. These activities include applied research, development tests, as well as evaluation of missiles, and other types of aerial weapons. Facilities available at this installation include specialized ranges, with instrumentation, and data reduction capabilities. Manned aircraft, range crews, and shop and laboratory space are available. These facilities can be utilized for the flight testing of armed missiles against ground, sea, or aerial targets to determine missile system effectiveness.

6-5.10. Air Proving Ground Center, Eglin Air Force Base, Florida

The Air Proving Ground Center is a facility for the engineering testing of air armament. An important and continuing part of the work at this center is to provide testing support to contractors and other Government agencies. To accomplish this, the Center is developing versatile test facilities and rapid methods of data reduction. Various types of air-to-ground ranges are available. The Center has a large and growing inventory of instruments with which to meet test requirements. Aircraft with their crews and maintenance personnel, range crews, instrumentation, and shop and laboratory space can be provided. Range facilities are available at the Center for firing tests of the smaller warheads, primarily those used in short range ground-to-air, air-to-air, and air-to-ground missiles. Range facilities are also available for a limited testing of cluster warheads. Parameters which can be investigated include warhead system functioning, bomblet dispersion and bomblet flight characteristics. In addition, facilities are also available for BW and CW testing.

6-5.11. Air Force Flight Test Center, Edwards Air Force Base, California

Though the activities of the Air Force Flight Test Center are not directly related to the development of missile warheads, there are two track test facilities available at this Center which may interest designers of such warheads. One of these tracks is 2,000 feet long, and the other 10,000 feet. It is suggested that the designer contact the Flight Test Center for specific information concerning the use of these tracks if dynamic testing of warheads is required.

6-5.12. Holloman Air Development Center, Alamogordo, New Mexico

The Holloman Air Development Center conducts research and development of guided missile systems and components; conducts tests and evaluations of missile weapon systems, missile operational techniques and associated equipment; and also conducts aeromedical research and development. Extensive range facilities are available at this Center for the flight testing of guided missile systems and the developmental testing of missile warheads. Armed missiles of relatively short ranges can be fired ground-to-ground, air-to-ground, air-to-air, or ground-to-air to determine their effectiveness against typical targets. Range facilities are also available for the various tests which are required in the development of cluster warheads. The Center also has a high speed track and support facilities such as manned aircraft, maintenance personnel, and instrumentation.

6-6. BIBLIOGRAPHY

(1) "Facilities of the United States Army for Research and Development, Volumes 1, 2, and 3". Office of the Assistant Chief of Staff, G-4 Logistics - Research and Development Division, Department of the Army, July, 1953.

(2) "California Naval Research, Devellopment and Test Stations - Unique Facilities", Interlaboratory Committee on Facilities, May, 1953.

(3) "Brochure of General and Descriptive Information of the U.S. Naval Proving Ground", July, 1949, N.P.G. Report No. 303.

(4) "Naval Aviation Ordnance Test Station, Descriptive Brochure", Feb. 1956.

(5) "Holloman Air Development Center - Information Guide", HQ HADC, Dec. 1956.

(6) "Air Force Armament Center - Range Facilities", HQ AFAC, January, 1954.

APPENDIX
CHARACTERISTICS OF HIGH EXPLOSIVES FOR MISSILE WARHEADS

Haller, Raymond & Brown, Inc., in an Appendix A to their report number 91-R-5, entitled "Survey of Guided Missile Warheads", included a summary of explosives for guided missile warheads. This is presented on the following pages, essentially in its entirety, for ready reference. It is to be noted that additional explosives have been developed or are being tested for use in guided missile applications since the publication of the survey.

A - INTRODUCTION

Six different explosive mixtures are used in the missile warheads included in this Survey. These are: RDX Composition B, H-6, HBX-1, Tritonal, RDX Composition C-3, and Cyclotol. Each of these explosives is a cast mixture (none are pressed mixtures) and has certain measurable properties which distinguish it from other explosives. The distinguishing properties of explosives are listed in Tables 1 and 2, at end of appendix.

The warhead operation determines the type of explosive that is used. Warheads considered are of the following types: fragmenting, continuous rod (expanding), blast, fragmentation-blast, and armor piercing. Some warheads use charges (called "propellants") which disperse submissiles or flechettes from the warhead at relatively slow rates to produce a controlled pattern. Propellants, however, will not be discussed.

There are four general requirements for all explosives; i.e., the explosive must be fluid enough in the preparatory state to be cast in the warhead, withstand shipping and handling, withstand the effects of time, and perform predictably when used.

Most of the production, handling, and time factors may be considered under the following classifications: fluidity (capability of being cast, etc.), shrinkage, fragility, stability and exudation (bleeding-out of components). Presumably, explosives used in the missile warheads in this Survey are satisfactory in production, handling and time factors - - otherwise they would not have been used in the warheads. These factors are not discussed in this report.

The topics to be discussed are the relative characteristics of several components, the tests used to derive some of the relative properties of explosives, and a comparison of properties of cast explosives used in the warheads surveyed.

B - EXPLOSIVE COMPONENTS

TNT is used as an index base for determining the relative characteristics of the two explosive components given in Table 1. TNT is used as a component in all six of the explosives cited in the Survey and comprises 4 to 80 parts (by weight) of the explosive mixtures. The heat of combustion TNT is greater than RDX. The Gurney velocity constant ($\sqrt{2E}$) of TNT is 6940 feet per second. TNT has a good blast effect (4*). Its detonation rate is 6745 meters per second.

The most common component used in the explosives in this Survey is RDX. In four of the six explosives discussed, RDX comprises 40 parts or more by weight of the mixture. RDX by itself is very difficult to cast (4). It has a low heat of combustion (0.63) relative to TNT (1.00), and a very high Gurney velocity constant, 8040 feet per second, or an index of 1.16.

* See references by number at end of Appendix.

Four explosive mixtures in this Survey contain either wax or D-2 desensitizer. Wax enhances some of the physical (handling) properties of the explosives mixtures since it acts as a desensitizer (4). D-2 improves the toughness of these mixtures but decreases stability in storage (4).

Aluminum is used in three of the six explosive mixtures. This enhances the flash effect of explosives and promotes a more controlled expansion during combustion. The addition of aluminum may reduce the fragility of the explosive and also minimize shrinkage during curing after the block of the explosive has been cast.

Nitrocellulose, tetryl, MNT, and DNT are used by only one of the explosives——RDX Composition C-3. No information is given in the Survey about these components.

C - EXPLOSIVE TESTS

Several measurements comparing the relative properties of explosive mixtures are shown in Tables 1 and 2. The tests from which the measurements are derived are discussed in order of appearance in the tables.

1. Peak Pressure TNT Equivalent Test

The test for the peak pressure equivalent of TNT compares the pressure produced by a sample explosive with that produced by an equal weight of TNT. The tests are made at the same standard distance.

The experimental set-up usually consists of a piezo-electric gauge located a standard distance from the center of the explosion. This gauge indicates the pressure impulse by voltage wave form when struck by the shock wave (7). Peak pressure is the maximum ordinate of the pressure-time curve determined from the experiment.

2. Positive Impulse Test

The test for TNT equivalent in positive impulse is identical to the test for peak pressure. Positive impulse is equal to the area under the pressure-time curve lying above the atmospheric pressure (7).

3. Ballistic Mortar Test

The ballistic mortar test (6) determines the weight of an explosive required to raise a heavy ballistic mortar the same height to which it is raised by 10 grams of TNT. The weight of explosive meeting this requirement is then used to compute its TNT equivalence by the formula

$$\text{TNT Value} = \frac{10 \text{ grams TNT}}{\text{Sample Weight in Grams}} \quad *$$

The physical set-up for this test consists of a heavy ballistic mortar suspended on a compound pendulum. The mortar contains a chamber about 6 inches in diameter and 1 foot long. A standard projectile occupies about 7 inches of this chamber, while the sample being tested occupies only a small portion of the remainder of the chamber. Upon detonation, the projectile is driven into a sand bank and the mortar swings through an arc. Swing height is recorded by a pencil attached to the pendulum.

4. Trauzl Test

The Trauzl test (6) determines the weight of an explosive required to cause the same expansion in a standard experimental measuring device as does TNT. Equivalent weights for the explosives tested are readily determined from this measurement and expressed as a ratio by the equivalent weight of TNT per unit weight of the test explosive.

The experimental set-up (6) uses desilverized lead cylinders 200 millimeters in diameter and 200 millimeters in height. In the end of each of these is centered a cavity 25 millimeters in diameter and 125 millimeters deep. A trial and error process is used to determine an amount of the explosive which will expand the cavity on detonation between 250 and 300 cubic centimeters. It has been found by Naoum that within this range of volume there is linear correlation between volume increase and sample weight.

5. Plate Dent Test

The brisance or shattering effect is

* Some of the references use percentages instead of fractional values.

determined by a plate dent test which measures the depth of the dents in a steel plate made by detonating equal weights of TNT and the explosive being tested. The measurements are used according to the formula

$$\text{Relative Brisance} = \frac{\text{Sample Dent Depth}}{\text{Dent Depth for TNT at 1.61 gm/cc}}$$

6. Detonation Rate Test

The detonation rate test measures the time-distance burning rate of a long piece of the explosive being tested. The index is determined from

$$I_{DR} = \frac{\text{DR for test explosive}}{\text{DR for TNT}}$$

The experimental set-up uses a rotating drum camera to record the burning rate of the explosive. The explosive is 1 inch in diameter and 20 inches long, and is held in place by a cellulose acetate sheet. A standard initiating system is used, consisting of four tetryl pellets at one end of the wrapped explosive in conjunction with a Special Corps of Engineers Blasting Cap placed in a central hole in the end pellet.

7. Gurney Velocity Constant Test

The Gurney velocity constant (8), $\sqrt{2E}$, is determined empirically from measurements of the average velocity of fragments from steel cylinders and other casing shapes. The constant is obtained directly from the formula

$$V_o = \sqrt{\frac{2E\,(c/m)}{1 + (1/2)(c/m)}}$$

where:
- V_o = the initial velocity of the fragment in feet per second,
- c = the weight of the explosive charge in grams,
- m = the weight of the fragmenting casing in grams, and
- E = a constant depending on the explosive measured in calories per gram.

The index is determined from

$$I\sqrt{2E} = \frac{\sqrt{2E}\ (\text{test explosive})}{\sqrt{2E}\ (\text{TNT})}$$

The experimental set-up for obtaining the Gurney velocity constant by determining initial fragment velocity usually makes use of a moving picture camera (7). One such set-up involves a high speed time-synchronized camera, a visual method (tetryl cap) for recording initiation of the explosion, and a material (duralumin) which flashes when struck by a fragment. This flash indicates the termination of fragment flight to the camera. Since the fragment travel distance is known, the initial velocity is readily computed (8).

8. Heat of Combustion Test

The heat of combustion test determines the heat content of equal weights of explosive mixtures. The measurements are made by calorimeters and are given in calories per gram.

9. Standard Cylinder Tests

The results shown in Table 2 were taken from two tests (1). Information on the number and mass of the fragments was obtained from pit tests. The velocity information was determined in a velocity range.

In the pit test (1), a cylinder is buried in a pit filled with sawdust. The cylinder is enclosed in a cardboard box so that the initial expansion takes place in air. The fragments are removed from the sawdust by magnets and by sifting.

The velocity range (1) uses a rotating drum camera which photographically records the passage of fragments past three illuminated vertical slits. Range geometry and the time lapse between the two outside slits determine the average fragment velocity. One-tenth of the cylinder circumference comprises the sample beam in which velocities are measured.

D - EXPLOSIVE MIXTURES

1. RDX Composition B (60/40/1: RDX/TNT/Wax Added)

RDX Composition B is the explosive

most commonly used in the guided missile warheads included in the Survey. As shown in Table 3 all but six of the twenty-five different type warheads containing this explosive are the fragmenting type. Five non-fragmenting warheads are continuous rod types. Alternatively, these five warheads may use the explosive, H-6.

RDX Composition B consists of three ingredients: RDX, TNT, and Wax. The notation in the above and subsequent captions gives the parts by weight of the components. Blending these ingredients provides an easily cast explosive which does well as a fragmenting and expanding (continuous rod) charge.

As noted in Table 1, the brisance of RDX Composition B is 1.32, the highest for the explosives found in the literature. In other words, this explosive is the most shattering. The high Gurney velocity constant (7610 feet per second) or index 1.10, is exceeded only by Cyclotol and H-6 among the explosive mixtures and indicates a high initial fragment velocity. The relative blast effect (heat of combustion) of Composition B is low, 0.78, compared with 1.00 for TNT. Blast, however, is not a primary requisite for a fragmenting warhead. The detonation rate of RDX Composition B is the highest for explosives found in the available literature; 7840 meters per second with an index of 1.17. Also, according to the Trauzl test (6) of volume expansion RDX Composition B was the highest, having an index of 1.30. The Ballistic Mortar test (6), which measures the relative energy of an explosive, gives an index of 1.33 for RDX Composition B.

RDX Composition B would also appear to be good as a fragmenting explosive from the relative measurements given in Table 2, which were made using standard cylinders as the fragmenting casings (1). This explosive exceeds H-6 and HBX-1 in both the number of fragments formed and the average initial velocity.

2. H-6 (74/21/5/0.5: Composition B/Al/ D-2 Desens./CaCl)

The explosive H-6 is used in ten different warheads in the Survey (Table 3). Of the ten warheads, five use RDX Composition B as an alternative explosive.

H-6 is closely similar to RDX Composition B. The principal difference in composition is a reduction in the proportion of RDX and TNT and the addition of aluminum to that mixture.

The equivalent weight of H-6 to TNT for peak pressure is 1.27, the highest attained among the explosives listed in Table 1. H-6 also ranks highest in the TNT equivalent weight for positive impulse, with an index of 1.38.

The Gurney Velocity Constant, $\sqrt{2E}$, is 7710 feet per second for H-6, an index of 1.11. Only Cyclotol and RDX Composition C-3, of the feasible compositions noted, surpass H-6 in ability to impart a high initial velocity to fragments. In a number of fragment blast warheads, H-6 (or HBX-1) provides a worthwhile increase in blast damage over RDX Composition B without a commensurate loss of damage from fragments.

3. HBX-1 (67/11/17/5/0.5: Comp. B/ TNT/Al/D-2 Desens./CaCl)

HBX-1 is used by three different warheads (Table 3), two of which are the blast type and one of the fragment type.

The TNT equivalent weight positive impulse for HBX-1 is 1.21, which is less than that for H-6, but greater than that for RDX Composition B. The same relationship holds for the TNT equivalent weight peak pressure. It would appear that H-6 is as good or better than HBX-1 for blast effect (heat of combustion measurement), having an index 1.06 compared with 1.03. As noted in Table 2, HBX-1 produces a larger number of fragments than H-6 and a consequent smaller average fragment mass when tested in a standard fragmenting cylinder. Average initial velocity, however, is lower than that for H-6. Lower velocity for HBX-1 is further exemplified by the difference in the Gurney velocity constant. Only Tritonal and TNT, have lower $\sqrt{2E}$ values.

4. Tritonal (80/20: TNT/Al)

The explosive Tritonal is used by a blast warhead for the Matador and a combination fragmentation-blast warhead for the Corporal missile.

As noted in Table 1, the TNT equivalent weight of peak pressure and impulse for Tritonal are 1.07 and 1.11, respectively. These measurements rank low relative to those for the explosives discussed previously. Most of the other measurable properties of Tritonal are also somewhat lower relative to those for TNT.

The heat of combustion for Tritonal is 1.21, highest among the explosives for which there is information in Table 1. This is a measure of the high blast potential of Tritonal. The brisance measure, 0.93, indicates a low shattering effect. This explosive is used when large fragment masses are required.

5. RDX Composition C-3 (77/4/1/3/5/10: RDX/TNT/Nitrocellulose/Tetryl/MNT/DNT)

RDX Composition C-3, is used in fragmentation warheads for the Honest John and Corporal missiles.

As noted in Table 1, RDX Composition C-3 ranks lower than RDX Composition B in the ballistic mortar test. It ranks lowest among the noted explosives in the Trauzl test. However in brisance, it ranks considerably above Tritonal. This is partially substantiated by the high detonation rate, 7625 meters per second (index 1.13), second only to that of RDX Composition B, 7840 meters per second (index 1.17). It has the highest $\sqrt{2E}$ value, 8800 feet per second (index 1.27). These ratings explain its use as a fragmenting explosive.

6. Cyclotol (75/25: RDX/TNT)

The explosive Cyclotol is utilized in the French SS-10 missile which uses a shaped charge type warhead against tanks. The present standard composition of Cyclotol is as noted above, but the composition utilized in this missile is 50/50: RDX/TNT.

The only measure available for Cyclotol is the Gurney velocity constant, $\sqrt{2E}$. This is 7850 feet per second, with an index of 1.13 lower only than RDX and RDX Composition C-3. This infers that Cyclotol has a very high peak pressure and positive impulse rating, as does H-6 which also has a high $\sqrt{2E}$ value.

E - CONCLUSIONS

An overall view of the following two tables, 1 and 2, enables one to make a direct comparison among the different explosives given in each measurement category. Unfortunately, some of the data were not available in the literature at the time of this survey. There is no absolute correlation of explosives by type of use or by rank in any of the measurement categories; however, the correlation of combinations of one or more uses and the measurements have been noted.

Changes of explosives used in the missiles in Table 3 are to be expected.

TABLE 1

PROPERTIES OF CAST EXPLOSIVE MIXTURES RELATIVE TO TNT

Explosive Mixtures	TNT Equivalent Peak Pressure (5)	TNT Equivalent Positive Impulse (5)	Ballistic Mortar	Trauzl (1)	Brisance (1)	Detonation Rate Index	Detonation Rate m/sec	Gurney Velocity Constant 2E (5) Index	Gurney Velocity Constant 2E (5) ft/sec	Heat of Combustion Rel. to TNT (Blast) (5)
RDX Comp. B	1.13	1.06	1.33	1.30	1.32	1.17	(7840)	1.10	(7610)	0.78
H-6	1.27	1.38						1.11	(7710)	1.06
HBX-1	1.21	1.21						1.05	(7260)	1.03
Tritonal 80/20	1.07	1.11	1.24	1.25	0.93	0.96	(6475)	0.91	(6280)	1.21
RDX Comp. C-3			1.26	1.17	1.18	1.13	(7625)	1.27	(8800)+	
Cyclotol 50/50								1.13	(7850)	
TNT	1.00	1.00	1.00	1.00	1.00	1.00	(6745)	1.00	(6940)	1.00
RDX								1.16	(8040)	0.63
Torpex			1.38	1.64	1.20	1.11	(7495)	1.07	(7450)	

(1) and (5) See References by number.
+ Picatinny Arsenal Technical Report No. 1530

TABLE 2

MEASUREMENTS OF EXPLOSIVE MIXTURES FROM STANDARD CYLINDER TEST FIRINGS (1)

Explosive	C/M	N_o	N(0.5) (gms.)	m(0.5) (gms.)	v_o ft/sec
Comp. B	0.378	3700	977	1.72	4440
H-6	0.385	2279	788	2.23	4420
HBX-1	0.386	2785	873	1.97	4130
Tritonal 80/20					
Comp. C-3					
Cyclotol 50/50					
TNT	0.358	1852	723	2.52	3710
RDX					
Torpex					

(1) Standard cylinder employed for fragmentation. Fragment distribution following Motts Law, N_o total no. of fragments, N(0.5 gm) no. of fragments over 0.5 gram m(0.5 gm) mean mass of fragments over 0.5 grams

TABLE 3
WARHEADS BY EXPLOSIVE USED

RDX Composition B					
Missile	Warhead	Type	Missile	Warhead	Type
Bomarc	T33	Frag.	Talos (6b)	Ex 17 Mod 1	Frag.
Corporal	T25E1	Frag.	*Tartar	Ex 20 Mod 1	Cont. Rod
LaCrosse	T34E1	Sh. Charge	Terrier	Ex 12 Mod 0	Frag.
Meteor	EX 8 Mod 0	Frag.	*Terrier Adv.	Ex 19	Cont. Rod
Nike I	T22E4	Frag.			
Nike I	T26E4	Frag.	H-6		
Nike I	T37	Frag.	Missile	Warhead	Type
Nike I	T37E2	Frag.	*Sparrow III	Ex 21 Mod 1	Cont. Rod
Nike I	T37E3	Frag.	*Sparrow 1A	Ex 22 Mod 1	Cont. Rod
Nike I	T38	Frag.	Talos (6a)	Ex 7 Mod 1	Frag.
Nike I	T38E2	Frag.	*Tartar	Ex 20 Mod 1	Cont. Rod
Nike I	T38E3	Frag.	*Terrier Adv.	Ex 19	Cont. Rod
Sparrow I	Ex 1 Mod 0	Frag.	*Terrier	Mk 5 Mod 3 and 6	Frag.
Sparrow III	Ex 2	Frag.	Falcon	GAR 3 Mod 0	Blast
Sparrow III	Ex 3 Mod 1	Frag.	Nike Hercules	T45 Fra	Frag.-Blast
Sparrow II	Ex 5 Mod 1	Frag.	Nike Hercules	T46	Cluster
*Sparrow III	Ex 21 Mod 1	Cont. Rod	Hawk	XM5	Frag.-Blast
*Sparrow 1A	Ex 22 Mod 1	Cont. Rod			
Sparrow 1	Mk 7 Mod 0	Frag.			
Talos (6a)	Ex 6 Mod 1	Frag.			
Talos (6b)	Ex 14 Mod 0	Cont. Rod			

*Indicates that this same missile warhead may also use another explosive

TABLE 3 (cont'd)
WARHEADS BY EXPLOSIVE USED

HBX-1			RDX-Composition C-3		
Missile	Warhead	Type	Missile	Warhead	Type
Bomarc	Cluster	Blast	Corporal	T39E3	Frag.
Honest John	T2021	Blast	Honest John	T39E3	Frag.
Sidewinder	Mk 8 Mod 0	Frag.	50/50 Cyclotol		
Tritonal			Missile	Warhead	Type
Missile	Warhead	Type	SS-10	French	Shaped Charge
*Corporal	T23E1	Frag. Blast			
Matador	T3E3	Blast			

*Indicates that this same missile warhead may also use another explosive

REFERENCES

1. Salem, A. D., Shapiro, N., Singleton, B. N., Jr.: *Explosives Comparison for Fragmentation Effectiveness*, NAVORD Report No. 2933, August 1953, Naval Ordnance Laboratory, White Oak, Maryland, Confidential.

2. Hyndman, J. R., Shuey, H. M., Iwanciow, B. L., Taylor, E. A.: *Quarterly Progress Report on Interior Ballistics*, Redstone Arsenal Report No. P-54-3, 10 May 1954, Redstone Arsenal, Huntsville, Alabama, Confidential.

3. Holt, P. L., McGill, R.: *Explosives Research Department Publications*, NAVORD Report No. 4163, 1 November 1955, Bureau of Ordnance, Department of Navy, Washington D. C., Confidential.

4. Anonymous: *Fundamental Development of High Explosives*, Progress Report No. 113, 1 November to 30 November 1955, Arthur D. Little, Inc., Cambridge, Massachusetts, Confidential.

5. Christian, E. A., Fisher, E. M.: *Explosion Effects Data Sheets*, NAVORD Report No. 2986, 14 June 1955, Naval Ordnance Laboratory, White Oak, Maryland, Confidential, AD-69244.

6. Tomlinson, W. R., and revised by Shetfield, O. E.: *Properties of Explosives of Military Interest*, Picatinny Arsenal Technical Report No. 1740 (U), April 1958, and Supplement No. 1 (C), August 1958, Picatinny Arsenal, Dover, N.J. Confidential.

7. Sachs, R. G., Bidelman, W. P.: *Blast Measurements on Five-and Ten-Ton Bare Charges of TNT*, Report No. 454, 1 March 1944, Ballistics Research Laboratories, Aberdeen Proving Ground, Maryland, Confidential, AD-73234.

8. Davids, N., Kohler, M. R., Jr., Duncan, R. L.: *Basic Compendium of the Fragmentation Properties of Fin-Stabilized Mortar Shells Including Theories, Data and Test Methods*, 15 April 1956, Haller, Raymond & Brown, Inc., State College, Pa., Confidential.

INDEX

Aberdeen Proving Ground - Test Facilities	185
Agents - Chemical and Biological Warheads	133
Air Force Flight Test Center - Test Facilities	188
Air Proving Ground Center - Test Facilities	188
Analytical Evaluation Method	154
Army Chemical Center - Test Facilities	186
Beam Width - Fragmentation Warheads	47
Fragment Patterns	50
Fragment Slow-Down	49
Fragment Velocity	47
Selection	52
Bibliography	
Blast Warheads	46
Chemical and Biological Warheads	138
Cluster Warheads	111
Continuous Rod Warheads	92
Discrete Rod Warheads	84
Explosives	198
Fragmentation Warheads	84
Shaped Charge Warheads	130
Warhead Evaluation	163
Warhead Selection	35
Warhead Testing	188
Blast Method of Ejection - Cluster Warheads	101
Blast Warheads	36
Bibliography	46
Detail Design Data	36
Design Data, Summary of	45
Explosive Loading and Sealing	41
Fabrication and Tooling	44
Function of Warhead Case and Related Blast Effects	36
Fuze Installation	44
Fuzing Requirements, Summary of	45
Installation and Handling Provisions	47
Strength Analysis	41
Weight and Space Allocated, Compatibility of	40
Detail Design Steps	36
External	36
Fundamental Concepts	149
Internal	36
References	45
Test Procedures and Techniques	182
Warhead Types	1
Bomblet Compartment and Structure - Chemical and Biological Warheads	130
Bomblets - Chemical and Biological Warheads	131

Characteristics of High Explosives for Missile Warheads	189
Characteristics of Service Warheads	138
Chemical and Biological Warheads	130
Agents	133
Bibliography	138
Cluster Type Warheads	130
Bomblet Compartment and Structure	130
Bomblets	131
Ejection Systems	132
Environmental Requirements	133
Massive Type Warheads	133
Warhead Types	11
Circular Normal Distribution - Distribution of Guidance Error	144
Classification of Targets - Warhead Selection	33
Cluster Submissiles - Test Procedures and Techniques	172
Cluster Type - Chemical and Biological Warheads	130
Cluster Warheads	92
Bibliography	111
Detail Design Data	92
Design Data, Summary of	110
Ejection Methods	100
Ejection System, Design of	103
Fuzing Requirements, Summary of	110
Number of Submissiles	100
Obstruction Removal Devices, Design of	109
Optimum Pattern	92
Retention System, Design of	109
Submissiles, Design of	105
Submissiles, Types of	100
Support Structures, Design of	107
Detail Design Steps	92
Fundamental Concepts	149
References	111
Test Procedures and Techniques	176
Warhead Types	10
Continuous Rod Warheads	87
Bibliography	92
Detail Design Data	88
Design Data, Summary of	92
Details, Warhead	98
Explosive Cavity	90
Explosive Charge	88
Fuzing Requirements, Summary of	98
Rod Bundle, Dimensions of	88
Rod Dimensions, Cross Sectional	88
Detail Design Steps	87
References	92

Warhead Types	8
Cost of the Contribution - Weapons System Concepts	31
Damage Classification - Evaluation Principles	146
Damage Mechanism Testing - Test Procedures and Techniques	170
Data Reduction and Interpretation	169, 182
Design Data, Detail	
Blast Warheads	36
Cluster Warheads	92
Continuous Rod Warheads	88
Discrete Rod Warheads	84
Fragmentation Warheads	46
Shaped Charge Warheads	123
Discrete Rod Warheads	84
Bibliography	87
Detail Design Data	84
Design Data, Summary of	87
Explosive Charge	85
Fuzing Requirements, Summary of	87
Rod Cross Section	85
Rod Length	84
Rod Material	85
Rod Velocity	85
Warhead Details	87
Detail Design Steps	84
References	87
Warhead Types	8
Distribution of Fuzing Error - Evaluation Principles	145
Distribution of Guidance Error - Evaluation Principles	141
Drag Coefficient	49
Dugway Proving Ground - Test Facilities	187
Ejection Methods - Cluster Warheads	100
Blast Types	101
Gun Types	101
Ejection Systems - Chemical and Biological Warheads	132
Blast Method	103
Gun Tube Method	102
Elliptical Normal Distribution - Distribution of Guidance Error	144
Environmental Requirements - Chemical and Biological Warheads	133
Evaluation Methods	154
Analytical	154
Geometrical Model	158
Graphical	158
Monte Carlo and Lotto	156

Overlay	157
Simulated	156
Evaluation Principles	**139**
Conditional Kill Probability	140
Damage Classification	146
Fuzing Error, Distribution of	145
Guidance Error, Distribution of	141
Overall Kill Probability	139
Evaluations, Approximate	**153**
External Blast Warhead	153
Fragmentation Warhead	154
Internal Blast Warhead	154
Evaluation, Warhead	**139**
Explosive Charge Design - Shaped Charge Warhead	**125**
Explosive Mixtures	**191**
Explosive Tests	**190**
Explosives for Missile Warheads, Characteristics	**189**
External Blast Warhead - Approximate Evaluations	**153**
Fragmentation Warheads	**46**
Bibliography	84
Detail Design Data	47
Beam Width	47
Charge to Metal Ratio (Actual) and Explosive Type	73
Charge to Metal Ratio (Maximum)	54
Design Data, Summary of	82
External Configuration	53
Fragmenting Metal, Design of	79
Fragment Shape and Material	73
Fragment Size Control, Methods of	74
Fragment Weight and Velocity, Optimum	57
Fuzing Requirements, Summary of	82
Static Fragment Velocity, Maximum Initial	56
Warhead Components, Design of	79
Detail Design Steps	46
Evaluation, Approximate	154
Fundamental Concepts	149
References	83
Test Procedures and Techniques	174
Warhead Types	3
Fragment Size Control, Methods of	**74**
Comparison of	77
Grooved Charge	76
Notched Rings	75
Notched Wire	76
Precut	74

(Other)	76
Fragment Weight and Velocity, Optimum	57
Aerial Target	57
Ground Target	65
Fragments - Test Procedures and Techniques	170
Fuzing Error, Distribution of - Evaluation Principles	145
Geometrical Model - Evaluation Methods	158
Glossary of Terms	viii
Graphical Evaluation Methods	158
Guidance Error, Distribution of - Evaluation Principles	141
Circular Normal Distribution	144
Elliptical Normal Distribution	144
Miss Distance	144
Rectangular Area, Probability of Hitting	143
Gun Type Method of Ejection - Cluster Warheads	101
Gurney Formulas	56
Holloman Air Development Center - Test Facilities	188
Incendiary Warheads	12
Inert and Exercise Warheads	13
Internal Blast Warheads - Approximate Evaluations	154
Kill Probability, Conditional - Evaluation Principles	140
Kill Probability, Overall - Evaluation Principles	139
Leaflet Warheads	12
Liner Design - Shaped Charge Warheads	123
Lotto - Evaluation Methods	156
Massive Type - Chemical and Biological Warheads	133
Miss Distance - Distribution of Guidance Error	141
Monte Carlo - Evaluation Methods	156
Naval Aviation Ordnance Test Station - Test Facilities	188
Naval Ordnance Laboratory - Test Facilities	187
Naval Ordnance Test Station - Test Facilities	187
Naval Proving Ground - Test Facilities	187

Optimum Pattern - Cluster Warheads	92
Overlay - Evaluation Methods	157
Picatinny Arsenal - Test Facilities	186
Planning of Test Program	167
Data Reduction and Interpretation	169
Establishing Test Program	168
Outlining Requirements	167
References	
Blast Warheads	45
Cluster Warheads	111
Continuous Rod Warheads	92
Discrete Rod Warheads	84
Fragmentation Warheads	83
Shaped Charge Warheads	129
Warhead Evaluation	158
Rod, Continuous, Warhead - See "Continuous Rod Warheads"	
Rod, Discrete, Warhead - See "Discrete Rod Warheads"	
Rods - Test Procedures and Techniques	172
Rod Warhead - Fundamental Concepts	152
Rod Warhead - Test Procedures and Techniques	181
Scope of Weapons System Concepts	30
Selection Chart, Warhead	34
Service Warheads, Characteristics	134
Shaped Charge Warheads	111
Bibliography	130
Detail Design Data	123
Design Data, Summary of	129
Explosive Charge Design	125
Fuzing Requirements, Summary of	129
Liner Design	123
Warhead Casing Design	126
Detail Design Steps	111
Fundamental Concepts	153
References	129
Test Procedures and Techniques	180
Warhead Types	10
Simulated - Evaluation Methods	156
Size of the Weapons System Design Team	31
Stabilized Submissiles - Cluster Warheads	106
Sterne Formulas	56
Submissiles, Design of - Cluster Warheads	105

Symbols, Definition of	xi

Targets, Classification of	33
Test Facilities	184
Aberdeen Proving Ground	185
Air Force Flight Test Center	188
Air Proving Ground	188
Army Chemical Center	186
Dugway Proving Ground	187
Holloman Air Development Center	188
Naval Aviation Ordnance Test Station	188
Naval Ordnance Laboratory	187
Naval Ordnance Test Station	187
Naval Proving Ground	187
Picatinny Arsenal	186
White Sands Missile Range	186
Testing, Warhead – Planning of Test Program	167
Test Procedures and Techniques	170
Damage Mechanism Testing	170
Cluster Submissiles	172
Fragments	170
Rods	172
Warhead and Warhead Components	174
Blast Warheads	182
Cluster Warheads	176
Fragmentation Warheads	174
Rod Warheads	181
Shaped Charge Warheads	180
Test Program, Planning	167
Data Reduction and Interpretation	169
Specific Program, Establishing	168
Test Requirements, Outlining	167
Tests, Explosive	190

Unstabilized Submissiles – Cluster Warheads	105

Warhead and Warhead Components – Test Procedures and Techniques	174
Warhead Casing Design – Shaped Charge Warheads	126
Warhead Detail Design – General	36
Warhead Evaluation	139
Bibliography	163
Evaluations, Approximate	153
Evaluation Methods	154
Evaluation Principles	139

Warhead Evaluation (Continued)	
Fundamental Concepts	149
References	158
Warhead Selection	33
Bibliography	35
Chart	34
Classification of Targets	33
Warhead Testing	167
Bibliography	188
Data Reduction and Interpretation	182
Facilities	184
Planning of Test Program	167
Procedures and Techniques	170
Warhead Types	
Blast Warheads	1
Chemical and Biological Warheads	11
Cluster Warheads	9
Continuous Rod Warheads	7
Discrete Rod Warheads	7
Fragmentation Warheads	3
Incendiary Warheads	12
Inert and Exercise Warheads	13
Leaflet Warheads	12
Shaped Charge Warheads	10
Weapons System Concepts	30
Applications to Warhead Design	32
Measure of Cost of the Contribution	31
Scope	30
White Sands Missile Range - Test Facilities	186

ENGINEERING DESIGN HANDBOOK SERIES

The Engineering Design Handbook Series is intended to provide a compilation of principles and fundamental data to supplement experience in assisting engineers in the evolution of new designs which will meet tactical and technical needs while also embodying satisfactory producibility and maintainability.

Listed below are the Handbooks which have been published or submitted for publication. Handbooks with publication dates prior to 1 August 1962 were published as 20-series Ordnance Corps pamphlets. AMC Circular 310-38, 19 July 1963, redesignated those publications as 706-series AMC pamphlets (i.e., ORDP 20-138 was redesignated AMCP 706-138). All new, reprinted, or revised Handbooks are being published as 706-series AMC pamphlets.

General and Miscellaneous Subjects

Number	Title
106	Elements of Armament Engineering, Part One, Sources of Energy
107	Elements of Armament Engineering, Part Two, Ballistics
108	Elements of Armament Engineering, Part Three, Weapon Systems and Components
110	Experimental Statistics, Section 1, Basic Concepts and Analysis of Measurement Data
111	Experimental Statistics, Section 2, Analysis of Enumerative and Classificatory Data
112	Experimental Statistics, Section 3, Planning and Analysis of Comparative Experiments
113	Experimental Statistics, Section 4, Special Topics
114	Experimental Statistics, Section 5, Tables
134	Maintenance Engineering Guide for Ordnance Design
135	Inventions, Patents, and Related Matters
136	Servomechanisms, Section 1, Theory
137	Servomechanisms, Section 2, Measurement and Signal Converters
138	Servomechanisms, Section 3, Amplification
139	Servomechanisms, Section 4, Power Elements and System Design
170(C)	Armor and Its Application to Vehicles (U)
252	Gun Tubes (Guns Series)
270	Propellant Actuated Devices
290(C)	Warheads--General (U)
331	Compensating Elements (Fire Control Series)
355	The Automotive Assembly (Automotive Series)

Ammunition and Explosives Series

Number	Title
175	Solid Propellants, Part One
176(C)	Solid Propellants, Part Two (U)
177	Properties of Explosives of Military Interest, Section 1
178(C)	Properties of Explosives of Military Interest, Section 2 (U)
210	Fuzes, General and Mechanical
211(C)	Fuzes, Proximity, Electrical, Part One (U)
212(S)	Fuzes, Proximity, Electrical, Part Two (U)
213(S)	Fuzes, Proximity, Electrical, Part Three (U)
214(S)	Fuzes, Proximity, Electrical, Part Four (U)
215(C)	Fuzes, Proximity, Electrical, Part Five (U)
244	Section 1, Artillery Ammunition--General, with Table of Contents, Glossary and Index for Series
245(C)	Section 2, Design for Terminal Effects (U)
246	Section 3, Design for Control of Flight Characteristics
247	Section 4, Design for Projection
248	Section 5, Inspection Aspects of Artillery Ammunition Design
249	Section 6, Manufacture of Metallic Components of Artillery Ammunition

Ballistic Missile Series

Number	Title
281(S-RD)	Weapon System Effectiveness (U)
282	Propulsion and Propellants
284(C)	Trajectories (U)
286	Structures

Ballistics Series

140	Trajectories, Differential Effects, and Data for Projectiles
160(S)	Elements of Terminal Ballistics, Part One, Introduction, Kill Mechanisms, and Vulnerability (U)
161(S)	Elements of Terminal Ballistics, Part Two, Collection and Analysis of Data Concerning Targets (U)
162(S-RD)	Elements of Terminal Ballistics, Part Three, Application to Missile and Space Targets (U)

Carriages and Mounts Series

340	Carriages and Mounts--General
341	Cradles
342	Recoil Systems
343	Top Carriages
344	Bottom Carriages
345	Equilibrators
346	Elevating Mechanisms
347	Traversing Mechanisms

Materials Handbooks

301	Aluminum and Aluminum Alloys
302	Copper and Copper Alloys
303	Magnesium and Magnesium Alloys
305	Titanium and Titanium Alloys
306	Adhesives
307	Gasket Materials (Nonmetallic)
308	Glass
309	Plastics
310	Rubber and Rubber-Like Materials
311	Corrosion and Corrosion Protection of Metals

Military Pyrotechnics Series

186	Part Two, Safety, Procedures and Glossary
187	Part Three, Properties of Materials Used in Pyrotechnic Compositions

Surface-to-Air Missile Series

291	Part One, System Integration
292	Part Two, Weapon Control
293	Part Three, Computers
294(S)	Part Four, Missile Armament (U)
295(S)	Part Five, Countermeasures (U)
296	Part Six, Structures and Power Sources
297(S)	Part Seven, Sample Problem (U)

UNCLASSIFIED

FOR REFERENCE ONLY

FOR REFERENCE ONLY

UNCLASSIFIED

CONFIDENTIAL

www.ingramcontent.com/pod-product-compliance
Lightning Source LLC
Chambersburg PA
CBHW062101220526

45471CB00010B/3564